Healthcare Strategies and Planning for Social Inclusion and Development

Volume 1: Health for All - Challenges and Opportunities in Healthcare Management

Healthcare Strategies and Planning for Social Inclusion and Development

Volume 1: Health for All - Challenges and Opportunities in Healthcare Management

Basanta Kumara Behera
Former Director, Advanced Centre for Biotechnology,
Rohtak, India

Ram Prasad
Associate Professor, Department of Botany
Mahatma Gandhi Central University, Bihar, INDIA

Shyambhavee
Senior Resident Physician, Department of Community Medicine
Atal Bihari Vajpayee Institute, New Delhi, India

Academic Press is an imprint of Elsevier
125 London Wall, London EC2Y 5AS, United Kingdom
525 B Street, Suite 1650, San Diego, CA 92101, United States
50 Hampshire Street, 5th Floor, Cambridge, MA 02139, United States
The Boulevard, Langford Lane, Kidlington, Oxford OX5 1GB, United Kingdom

Notices
Knowledge and best practice in this field are constantly changing. As new research and experience broaden our understanding, changes in research methods, professional practices, or medical treatment may become necessary.

Practitioners and researchers must always rely on their own experience and knowledge in evaluating and using any information, methods, compounds, or experiments described herein. In using such information or methods they should be mindful of their own safety and the safety of others, including parties for whom they have a professional responsibility.

To the fullest extent of the law, neither the Publisher nor the authors, contributors, or editors, assume any liability for any injury and/or damage to persons or property as a matter of products liability, negligence or otherwise, or from any use or operation of any methods, products, instructions, or ideas contained in the material herein.

Library of Congress Cataloging-in-Publication Data
A catalog record for this book is available from the Library of Congress

British Library Cataloguing-in-Publication Data
A catalogue record for this book is available from the British Library

ISBN 978-0-323-90446-9

For information on all Academic Press publications
visit our website at https://www.elsevier.com/books-and-journals

Publisher: Andre G. Wolff
Acquisitions Editor: Elizabeth Brown
Editorial Project Manager: Sam W. Young
Production Project Manager: Kiruthika Govindaraju
Cover Designer: Alan Studholme

Typeset by STRAIVE, India

Contents

About the authors

Basanta Kumara Behera (*Former Director, Advanced Center for Biotechnology, MDU Rohtak, India; Former Adviser, Sanmar Speciality Chemical Ltd., India*)

Prof. Basanta Kumara Behera was a professor of biotechnology at three distinguished Indian universities, where he had been regularly associated with teaching and research at postgraduate level courses on the topic related to bioenergy management and biomass processing technology. In 2009, he joined an MNS company as an adviser for speciality chemical production and drug design through microbial process technology. Prof. Behera is associated with reputed national and international companies as technical adviser for the production of biopharmaceuticals under cGMP norms. He has the credit of being authored books published/in process by CRC, Springer Verlag, Elsevier, CABI.

Ram Prasad (*Associate Professor, Department of Botany, Mahatma Gandhi Central University, Motihari, Bihar, India*)

Dr. Ram Prasad is associated with the Department of Botany, Mahatma Gandhi Central University, Motihari, Bihar, India. Previously, he was assistant professor at Amity University, India; visiting assistant professor, Whiting School of Engineering, Department of Mechanical Engineering at Johns Hopkins University, Baltimore, United States, and research associate professor at School of Environmental Science and Engineering, Sun Yat-Sen University, Guangzhou, China. His research interests include applied microbiology, plant microbe-interactions, sustainable agriculture, and nanobiotechnology. Dr. Prasad has more than 200 publications to his credit, including research papers, review articles, and book chapters, five patents issued or pending, and edited or authored several books. Dr. Prasad has 13 years of teaching experience, and has been credited several awards on national and international level.

Shyambhavee (*Department of Community Medicine, University College of Medical Sciences and GTB Hospital, New Delhi-110095, India*)

Dr. Shyambhavee is a medical graduate from Lady Hardinge Medical College, New Delhi, and MD degree holder from University College of Medical Science, New Delhi. She has been working in the field of noncommunicable disease, immunization, maternal and child health, epidemiology, and health administration to upgrade and bring amendments in community for sustainable life pattern with good health. She is currently working as Senior Residence, Department of Community Medicine, University College of Medical Sciences and GTB Hospital, New Delhi, India.

Preface

Conceptual development of "Health for All" is to provide personal state of well-being within the reach of everyone in a given country. Health for all should not be confused with the availability of health services at individual level, but to take overall personal state of well-being. For example, WHO supports a nation to achieve SDG goals while supporting "Health for All" core activities. This is because healthy people's families and communities act as pillars for strengthening a country in all respects. So, wide coverage is given to universal health and possible source of financial funding for those in need, regardless of their ability to pay. But, it is unfortunate that, currently, 1 out of 17 of the world's citizen is suffering from health problem due to lack of the access to essential health services. So, this piece of work covers how international organization like WHO supports national health authority in functioning their core health-care system by providing support to health workforce, service delivery, and health information system, ensuring basic facilities such as health coverage, primary health-care education, etc.

The books mainly deal with public health services that not only improve the health service but also bring economic reforms especially in rural community. The proposed book covers the entire spectrum of health and well-being.

First chapter explains how hospital community health centers and nursing homes provide preventive curative and rehabilitative care. In addition, it describes how security professions, fire, ambulance providers, and emergency medical services can be closely coordinated for increasing the efficacy of health services. The authors want to develop awareness among doctors to mobilize other health-care professions to manage the system under extreme climatic and disaster condition to save life and bring social stability.

The last part of the first chapter gives an insight on disease classification and under what circumstance a disease outbreak become epidemic and pandemic at global level, the authors try to explain it with the example of COVID-19 challenge and bring to the attention of public to either eradicate or bring stability with the passing of time. In this connection it is recommended to develop well-organized health-care system either on temporary basis or upgrade the existing health-care system to meet the demand during emergency.

The second chapter converses the socioeconomic significance of universal health coverage to bring a sustainable healthy life. In this connection various fund raising systems without pocket money expenditure are also explained with various examples currently being applied in different countries. In this connection the role of government and private bodies is also explained. In addition, WHO protocols on universal health coverage by promoting rehabilitations, providing sufficient accesses for effective service and emergency treatment, under financial hardship are presented. The overall strategies of UHC to provide everyone access to the services that address the most significant causes of diseases and death and ensure that the quality of those services is good enough to improve the health of the people who receive them have been explained to convince and bring awareness for healthy life in social front. The authors given emphasis how SDG3 of universal health reformation can reduce the death caused due to road injuries, negligence in sexual reproductive. In addition, it has been explained how family planning and primary health education bring social understanding on health care and cooperation with various regulatory bodies in management and in bringing sustainable community health coverage at global level. The targets of 13 health care services in SDG 3 health services to save the population from alarming condition of pandemic are explained. The last part of this chapter explains statistic report on world health strategies and need for integrated cooperative services to save life globally.

The third chapter explains how lack of public health-care coverage is responsible for causing poverty and food security and unstable social life. The authors mainly placed emphasis on prevention to stop the spread of epidemic. For example, frequent hand washing, regular bath, sensitization maintenance are primary responsible for changing epidemic into pandemic form. In this connection emphasis is placed on the rule and protocol for washing hands. In this connection special instructions are given to not touch eyes, nose, and mouth without washing hands. In addition, it is explained how touching of the surrounding and carpets commonly contaminated by aero-drop from contaminated person or patients can transmit disease. It has also been clear how noncommunicable diseases are also responsible for alarming life its timely care is not taken. It also explains how inadequate access to health infrastructure, communication gap, deficiency in medicine, and health aides are responsible for health service, timely. This chapter also covers the SDG3 goals responsible for bringing harmony in health-care system with international collaboration project.

Abbreviation

CBHI	Community-Based Health Insurance
CDCP	Centers for Disease Control and Prevention
CDSS	clinical decision support system
cGMP	current good manufacturing practice
CHA	community health assessment
CMS	Centers for Medicare Services
CRM	customer relationship management
CRT	controlled room temperature
EAHP	European Association of Hospital Pharmacists
EAP	European Action Plan
ECE	early childhood education
EDRM	emergency and disaster risk management framework
EOC	emergency operations center
EPHO	essential public health operation
EVD	Ebola virus disease
FDA	Food and Drug Administration
FIFO	first in first out
GCP	good clinical practice
GDP	good distribution practice
GDP	gross domestic product
GLP	good laboratory practice
GLSC	green logistics and supply chain
GTIN	Global Trade Item Number
HACCP	hazard analysis and critical control points
HER	electronic health record
HIT	health information technology
ICD	International Classification of Diseases
ICI	information and communication technology
ICMR	Indian Council of Medical Research
ISDR	International Strategy for Disaster Reduction
LIFO	last in first out
MDGs	Millennium Development Goals
MERS	Middle East respiratory syndrome
MMR	maternal mortality ratio
MoHFW	Ministry of Health and Family Welfare
MPI	master patient index
MRO	maintenance, repair, and operation
MRSA	methicillin-resistant *Staphylococcus aureus*
MSF	Medicines Sans Frontiers

NCDC National Center for Disease Control
NGOs Nongovernment Organizations
NHS National Health System
OECD Organization for Economic Cooperation and Development
PHCs Public Health-care Center
QA quality assurance
QC quality control
RMNCAH reproductive maternal newborn child and adolescence health
RPMs remote patent monitoring systems
RTVN real-time value network
RUPRI The Rural Policy Research Institute
SAMM safety assessment of marketed medicines
SARS severe acute respiratory syndrome
SCM supply chain management
SCMSP supply chain management service provider
SDGs sustainable development goals
SDH social determinants health
SDOH social determinants of health
SHI Social Health Insurance
SHP social health protection
SSA Social Security Administration
TQM total quality management
UDI unique device identification
UHC Universal Health Coverage
UN United Nations
UNICEF United Nations International Children's Emergency Fund
USAID United States Agency for International Development
WBG World Bank Group's
WHO World Health Organization
WONCA World Organization of National College and Academy

CHAPTER

Public health and management 1

1.1 Introduction

Public health is a broad term applied to practice to strengthen health services for community well-being under which people can maintain good health, improve their health, or prevent the deterioration of their health. Public health covers the entire spectrum of health and well-being. Public health-care providers are mainly hospitals, physicians, community health centers, and nursing homes that provide preventive, curative, and rehabilitative care. Security professions, fire, ambulance providers, and emergency medical services are closely associated with public health providers.

Public health services function effectively with the collaborative efforts of multidisciplinary teams consisting of doctors, nurses, medical assistants, and many other. Public health management covers the administrative and managerial capacities, organizational structures, and systems needed to finance and deliver health services more efficiently, effectively, and equitably.

1.2 Definition

"Public health is the science governing with prevention of disease that is a threat to the overall status of health performance of a community, as well as with prolonging life and promoting health" [1]. The main functional aspect of public health is to put organized efforts for the well-being of the community with a healthy lifestyle. Delphi recommends facilitating technique to obtain anonymous acceptance of its suggestion for use in health research and public health development [2,3].

In 1997, Delphi study was able to develop a set of "essential public health functions" [4], which was at a later stage modified by the Pan America Health Organization and the World Health Organization (WHO) Regional Office for the West Pacific [5]. The Delphi method is a process used to arrive at a group opinion or decision by surveying a panel of experts.

For further adaptation, WHO Regional Office for Europe (WHO, EURO) has developed 10 Essential Public Health Operation (EPHO) [6].

1.3 The history of public health

In the beginning, the public health developers had to confront with some pessimistic people with vested interest, who used to discourage the public health services procedural follow-up by quoting some unusual social ethics. Opposition to Jennerian

Healthcare Strategies and Planning for Social Inclusion and Development. https://doi.org/10.1016/B978-0-323-90446-9.00001-0

vaccination, unnecessary criticism of the work of great pioneers in public health measures such as Louis Pasteur, Florence Nightingale, and many others had unnecessarily delayed the innovative breakthrough in preventing diseases.

Jenner's vaccination is the most interesting factual story on public health services. The Egyptian mummies from the 18th and 20th Egyptian Dynasties (1570–1085 BC) having small lesions resembling those of smallpox was the first evidence of smallpox existing. But it is believed that smallpox first appeared around 10,000 BC, at the time of the first agriculture settlements in Northeastern Africa. European states were victimized sometime between the fifth and 6th centuries and later were brought to the New World by Spanish and Portuguese conquistadors, where it decimated the native populations. Smallpox is mainly caused by exposure to the *Variola* virus and infection typically begins like a common cold.

Although opposition to Jenner's vaccination continued till the late 19th century in some areas, its supporters gradually gained ascendancy, ultimately leading to the global eradication of smallpox.

In 1861, Louis Pasteur published germ theory which proved that bacteria caused diseases. This idea was taken up by Robert Koch in Germany, who began to isolate the specific bacteria that caused particular diseases, such as TB and cholera. Louis Pasteur's germ theory was discarded at the initial stage. People disapproved Louis Pasteur's model of infectious diseases and argued that Antoine Bechamp's theory was right [7]. Another observation was that the diseased tissues attract germs rather than being caused by them [8]. This was also a contradicted statement against germ theory developed by Louis Pasteur.

In 1881, Louis Pasteur developed a vaccine for anthrax, which was used successfully in sheep, goats, and cows. In 1885, while doing research on rabies, Pasteur tested his first human vaccine. This vaccine was developed by attenuating the virus in rabbits and subsequently harvesting it in their spinal cord. After his great success in developing, rabies vaccine for humans Pasteur was able to conclude that if a vaccine could be found for smallpox, vaccines could also be found for all diseases. In 1878, Pasteur successfully cultured the bacteria responsible for causing chicken cholera and began inoculating chickens (Fig. 1.1).

But during such a trial many chickens died. So, at a later phase, he emphasized on safe inoculation methods in order to avoid chicken death during inoculation. The incidental story on the chicken cholera vaccine was quite interesting. In 1879, Pasteur noticed that old bacterial cultures were lacking the virulence. Incidentally, Pasteur gave instructions to laboratory assistance to inoculate fresh culture of the viral bacteria before leaving for holiday. Unfortunately, the laboratory assistant forgot to inoculate the chicken with fresh bacterial culture. After returning from 1 month holiday, he performed the same job by inoculating old culture to the chicken and surprisingly could notice only mild signs of diseases and chickens survived. At a later phase when the chicken was healthy, Pasteur inoculated the healthy chicken with fresh bacterial inoculums and surprisingly noticed the chicken with good health.

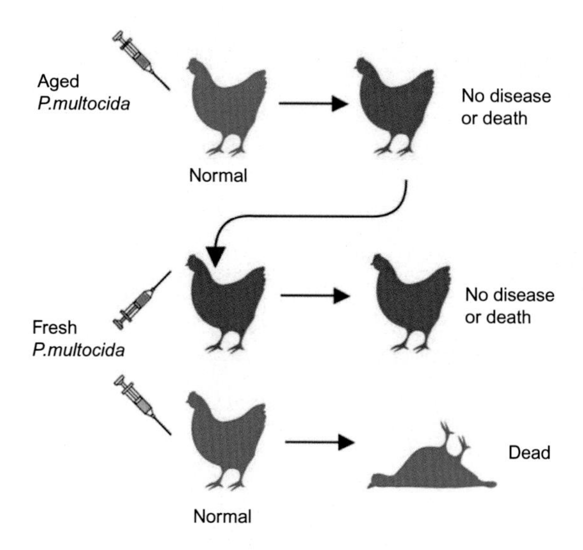

FIG. 1.1

Pasteur's fowl cholera experiment. Birds inoculated with an aged culture of *Pasteurella multocida* did not die. However, when subsequently inoculated with a fresh culture of virulent *P. multocida,* the birds were found to be protected. It was this experiment that launched the science of immunology.

Pasteur came to conclusion that by exposing to ambient atmospheric oxygen condition lost their virulence potential. The discovery of the chicken cholera vaccine by Louis Pasteur revolutionized work in infectious diseases and can be considered the birth of immunology.

The conceptual development of public health-care approaches can be well explained under the following heads.

1.3.1 Public health care and social ethics

Epidemics such as plague, cholera, and smallpox were dominated before the 18th century. Under the cloud of social ethics and inheritance of mythological believes people depend on God's blessing to prevent and cure diseases. People believed in a healthy deity (god or goddess) in mythology or religion associated with health, healing, and well-being, and also related to childbirth or mother goddesses (Fig. 1.2).

Epidemic diseases were used as signs of poor moral and spiritual conditions, to be governed through prayer and piety (Fig. 1.3). Some pandemic epidemic was monitored through isolation of ill and quarantine of travelers. In the late 17th century, several European cities appointed public authorities to adopt and enforce isolation and quarantine measures [9].

FIG. 1.2

People worship energy goddess for curing CORONA-19 (South China, India).

Hindu residents wearing protective masks perform prayers for protection against coronavirus outside a temple, in Ahmedabad, India [Amit Dave/Reuters]

Ahmed, 57 and his son, 10, perform Friday prayers in their home as mosques are closed due to concerns about the spread of coronavirus disease in Casablanca, Morocco [Youssef Boudlal/Reuters]

FIG. 1.3

Mass prayer for COVID-19 performed in different countries.

1.3.2 Restriction in public movement

The practice of isolation of ill, and quarantine of travelers before entering new countries were noticed to be a common measure for the prevention of pandemic contagious diseases. In 1701, Massachusetts passed the law for quarantine for traders and isolation of sick suffering from smallpox. By the end of 18th century, several cities like Boston, Philadelphia, New York, and Baltimore had well-framed quarantine and isolation rules as prevention measures against pandemic diseases [10].

During this period most of the developed cities established voluntary general hospitals for the physically ill and public institutions for the care of the mentally ill. Even the government made an official rule to take care of the physically and mentally ill by the neighbors in local communities [11,12].

By the end of the 18th century, several communities started demanding better formal arrangements for the care of their ill than Poor Law practices. The first American voluntary hospital was established in 1752 and in New York in 1771. The first public mental hospital was established in Williamsburg, Virginia in 1773 [13].

1.3.3 Sanitization awareness and public health care

Public awareness on sanitization intensified at the beginning of 19th century as an indicator of poor social and environmental conditions, and as well as for social and spiritual conditions.

In the 19th century, during the Victorian Era, the health and sanitization conditions were not to the mark. Tuberculosis, smallpox, measles-like bacteria/virus contamination in London was common. This was mainly due to horse transportation and cesspools filled the street with feces, and virus contamination was unavoidable (Fig. 1.4).

In the 19th century, horses were walking all day on the street, making the street road dirty and unhygienic with feces. The government employed young boys to clean the street road covered with feces from horses. But no improvement was noticed and

FIG. 1.4

London roads were dumped with horse's wastes (during 1800s). Near Regent street used to have 1000 horses for transport system.

FIG. 1.5

Georgian Era Toilets (Toilets in victorian-era.org).

the London Street was extremely unhygienic with horse feces. In Victorian era, every home used to have a cesspool (Fig. 1.5).

The only known place with toilets was called the Crystal Palace. Basically, cesspool was just deep holes that people would go to the bathroom in. Over time, the cesspool would fill up, and night workers would go around and empty them. It was illegal to empty the cesspool in the daytime due to an extremely foul smell. Some villagers would dump buckets of feces onto the sidewalk in the middle of the day which was another huge sanitary issue. In 1750, the population of Europe increased rapidly, with the simultaneous increase in infant death growth rate. These unfortunate incidents led to the rapid development of voluntary hospitals in the United Kingdom. The people's efforts had drawn the attention of the government to initiate precaution measures and extend public health facilities. In 1752, British physician Sir Johan Pringle published a book that explained ventilation in barracks and the provision of latrines.

In a developing country like India, sanitization has been a priority since ancient time, as scripted in Vedic times. The remnants of Indus Valley showed the awareness of people for public hygiene (Fig. 1.6).

But with the onset of the colonial role, sanitization ceased to be a national priority. The gradual increase in rural poverty under the colonial role has brought tremendous declination in rural hygiene conditions and ill health issues. During colonial rule, sanitization was given the least priority. In 1865, British Royal Commission reported a high mortality rate of 69 out of 1000 troops due to diarrhea. As a result, in

FIG. 1.6

The Harappan Bathroom (Harappa harappa.com).

1865, sanitization police were established under the Military Cantonments Act, and for the first time, sanitization boards were created in each province to look after civil sanitization conditions. This act was restricted to military areas rather than for the public. By 1947, India population was about 30 cores, had less than 1% sanitization coverage, and this statistical data prolonged for a long period.

Sanitization future prospects

The intensity of the COVID-19 pandemic has brought complete awareness among the people on personal hygiene. Things like regular and time-to-time hand washing (Fig. 1.7), use of face masks, maintaining a distance of 5 ft. seems to be casual, but

FIG. 1.7

Showing demonstration of hand washing as preventive measure for COVID-19.

are immensely important from a prevention point of view of contamination of infectious diseases like COVID-19.

Fabric masks are recommended to prevent onward transmission in the general population in public areas, particularly where distancing is not possible, and in areas of community transmission. So, masks are the physical barrier to respiratory droplets that may enter through the nose and mouth. The porosity of the fabric used for a mask is mainly dependent on the size of the pathogenic bacteria or virus (Fig. 1.8).

Nasal droplets and aerosols are more or less similar in size, less than 5 μm and can be transmitted over distances 1–2 m. The literature says "droplets" can be deposited as far away as 6–8 m. In the hospital environment, high infection rates were registered for several "aerosol-generating procedures" (invasive and noninvasive ventilation, intubation, tracheostomy, etc.); again, it has not yet been proved that aerosols are the culprits for the increased transmissibility but health-care workers are advised to wear N95/FFP2 or FFP3 in such a context. Currently, different types of face masks are available in the market to protect oneself from COVID19. The basic feature of the mask is as follows (Fig. 1.9A and B).

Mask provides against contaminants in the air, ranging from pollen to chemical fumes to pathogen. The quality of the mask depends on the nature of the fabric and design of the masks. Both disinfectants and sanitizers are used for controlling the infection. By means of sanitization, about 99.9% of surface pathogens are killed, whereas disinfections completely kill the pathogens. Disinfectants are made from quaternary ammonium compounds (quats), chlorine, (sodium hypochlorite, bleach), accelerated hydrogen peroxide or phenolic sanitizers. Sanitizers are chlorine, quats, iodine, and acid anionic. It is assumed that the use of hand sanitizers would become as normal in 2021 as sipping tea very often during winter.

On the 11th of February, WHO declared the name of the new diseases from the virus of Coronaviridae family in the Nidovirales order known as "COVID-19." The name of this virus is derived from Latin corona means crown-like structure on the outer side of the virus [14].

There are mainly three ways of contamination of coronaviruses, including COVID-19. These are the direct route, indirect contact, and airborne transmission.

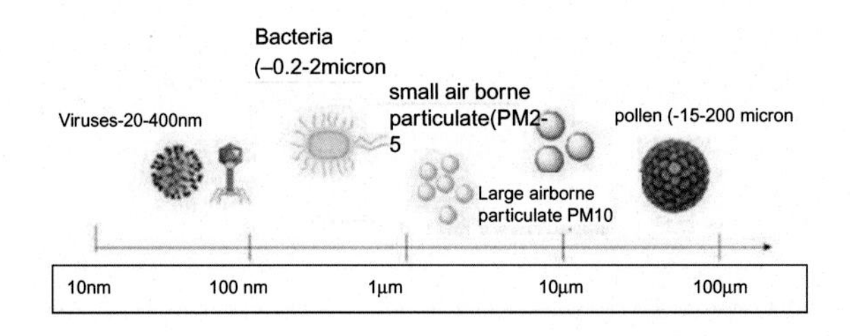

FIG. 1.8

Various size of bacteria and viruses.

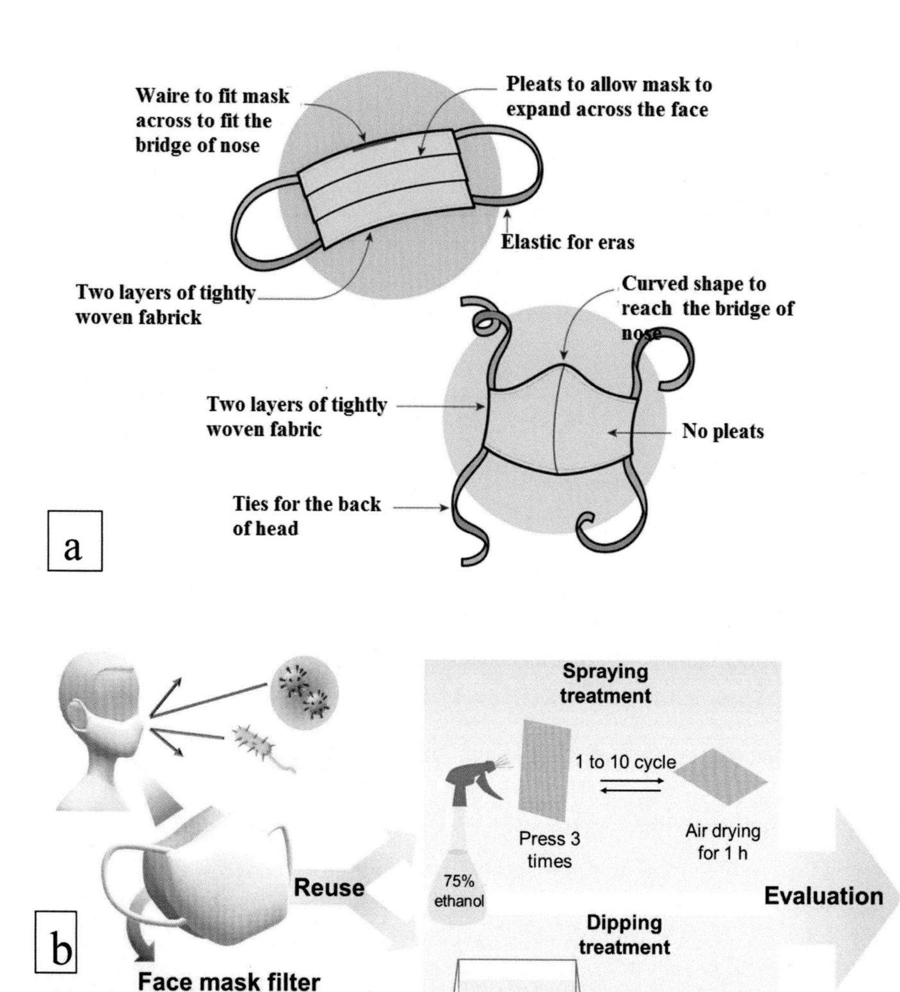

FIG. 1.9

Basic feature of mask (A) mask made from normal fabric, (B) mask made from nanopore material.

The epidemic (COVID-19) is transmitted from human to human by multiple means, namely by droplets, aerosols, and fomites (Fig. 1.10).

The direct mode of transmission is from an infected individual to a healthy person by close contact without any prevention measures like using the mask. While the indirect mode of transmission of the virus is due to touching or using objectives earlier used or touched by an infected person [15].

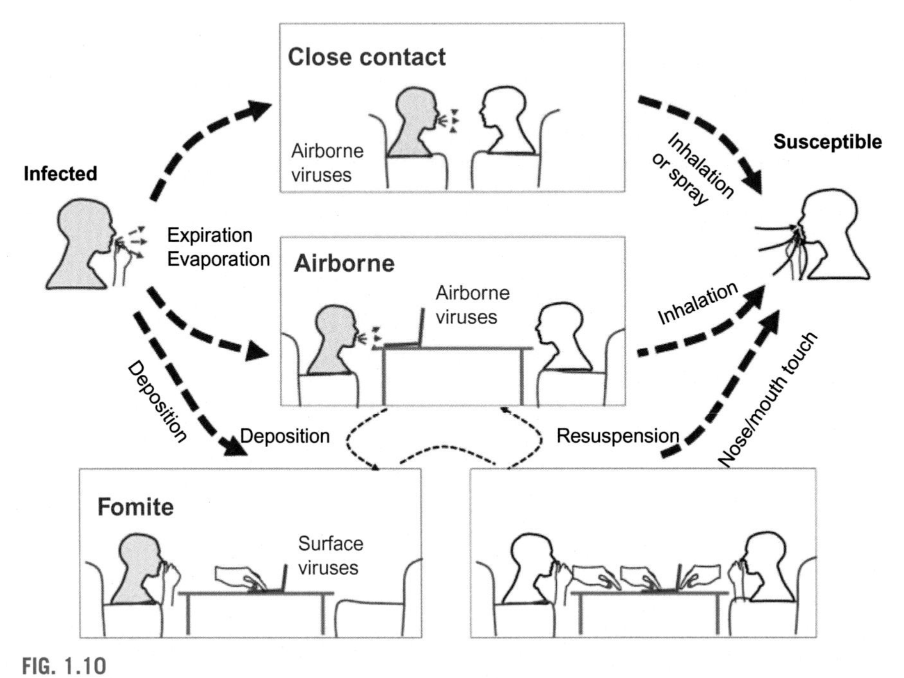

FIG. 1.10

Three ways of transmission of COVID-19 from contagious person to a healthy person.

1.3.4 Preventive measures

The end of 19th century was remarkable for rapid advances in scientific knowledge about the causes and prevention of numerous diseases caused by various pathogens. As stated earlier, in 1884, the concept of artificial immunization was developed by Louis Pasteur. During the following few years, the discovery of bacteriologic agents of diseases was made in European and American laboratories for coetaneous diseases such as tuberculosis, diphtheria, typhoid, and yellow fever [16].

During the same period lot of scientific data were available on the identification of bacteria and water purification techniques provided a means of controlling the spread of diseases and even preventing diseases. The basic concept of germ theory brought awareness about communicable diseases among the public and make the public understand that disease is mainly based on a single, specific cause. Scientists also revealed that both environment and people are sources of diseases. So, during this period, public agencies were more alerted to realize the significance of sanitization to avoid the contamination of diseases.

Identification of pathogens and interference of immunization and water purification techniques were used to control and prevent diseases. Scientists were immensely involved in research to understand the exact cause of disease and developed measures for controlling the disease.

For the first time, in 1890, the state and local health departments in the United States began to establish laboratories. The first was established in Massachusetts, as a cooperative venture between the State Board of Health and the Massachusetts Institute of Technology. In New York City, it was developed as a part of the New York City Health Department.

In 1891, W.T. Sedgwick, the most famous worker in sanitation and bacteriologic research, and consulting biologist for Massachusetts was able to identify the presence of fecal bacteria in water as the cause of typhoid fever and developed the first sewage treatment techniques. In 1890, Sedgwick conducted research on bacteria in milk and explained the benefit of pasteurization of milk.

In 1884–86, Theo Bald Smith, an American microbiologist, discovered the causes of several infectious and parasitic diseases. He found out that animals can be made immune to a disease by the injection of heat-killed cultures of the bacterium responsible for a specific disease. He was able to point out that the actual cause of hog cholera is a virus rather than *S. choleraesuis*, and developed techniques on the preparation of vaccine by a heat-killed method by disease-causing microorganisms. In 1888–93, Smith discovered Texas cattle fever caused by a protozoan parasite (*Pyrosoma bigeminum*, later named as *Babesia bigemina*) that is uninfected cattle by blood-sucking ticks.

1.3.5 The move toward personal care

In the early 20th century, both the local public and state government health departments became more conscious about public health. Identification and treatment of individual cases of diseases became the first priority. Massachusetts, Michigan, and New York City began producing and dispensing antitoxins in the 1890s. In 1907, Massachusetts Started reporting individual cases of 16 different diseases. At a later stage, cases of cancer also added in registering diseases.

Based on the data survey, it was realized that providing immunizations and treating diseases did not solve all health problems. Numerous diseases, such as tuberculosis, still remain unsolved despite sincere efforts. During World War I, it was understood that a substantial portion of the male population was either physically or mentally unfit for combat [17].

Overall results from registration also clear highest rates of morbidity persisting among children and poor. So, in order to have control over child mortality, the New York and Baltimore health departments started home visits by public nurses. To bring awareness among the people, New York health-care centers started a campaign on education on tuberculosis. In 1894, school health clinics were developed in Boston, New York. Subsequently, many cities like New York (1903), Rhode Island (1906) also developed child health-care centers. In 1915, more than 500 tuberculosis clinics and 538 baby clinics were developed in America [12].

Gradually, people have given to this movement a public health care for the prevention of diseases. Epidemiology provided a scientific justification for health programs that had originated with social reforms. Subsequently, in 20th century, scientists started research and development work on detailed analysis of diseases and treatment protocol. So, keeping in view such development, in 1923, C.E.A. Winslow defined public health as: "as Prolong life, and promoting physical health and efficiency" [10].

1.3.6 Toward improvement of public health

The late 19th century and early 20th century have witnessed federal involvement in public health development programs. In 1887, the National Hygiene Laboratory was established in the Marine Hospital in Staten Island, New York, in collaboration with chemistry, zoology, and pharmacology. In 1906, Congress passed the FOOD and Drug Act, for initiating manufacturing and, control, and sales of food and drugs. In 1912, the Marine Hospital Service was named the US Public Health Service. In 1914, Congress passed the Chamberlain-Kahn Act, to establish the US Interdepartmental Social Hygiene Board, a comprehensive venereal diseases control program for the military, and provided funds for quarantine of infected civilians. The Children's Bureau (founded in 1912) conducted the first White House Conference on child health in 1919 [10].

In 1922, the Sheppard-Towner Act was established by the Federal Board of Maternity and Infant Hygiene in order to provide administrative funds to the Children's Bureau to develop and conduct programs in maternal and child health. Through this act direct funding was provided for personal health services. The state government availed these funding facilities after the submission of a detailed report on the development plan and its proper implementation for improving or developing nursing facilities, home care, health education, and obstetric care mothers in the state; to designate a state agency to administer the program and to report on operations and expenditures of the program to the federal board [18].

Vaccines, which bridge the gap between biomedical science and public health, are one of the greatest achievements of the 20th century. Mass vaccination was the most important step taken as preventive measure for some deadly communicable diseases such as smallpox. Many methodological advances have facilitated a better understanding of diseases processes and opportunities for control.

Following are the few important developments in vaccination during 20th and 21st centuries (Table 1.1).

1.3.7 Finance involvement

Developed country like United States, spends more per capita health expenditure as compared to developing countries. Health spending per person in the United States was $ 10,966 in 2019, which was 42% higher than Switzerland, the country with the next higher per capita health spending (Fig. 1.11).

Table 1.1 Showing few important developments in vaccination during 20th and 21st centuries.

Year of development	Type of diseases	Name of scientist
1921	Tuberculosis	Albert Calmette
1923	Diphtheria	Gaston Ramon, Emil von Behring and Kitasato Shibasaburo
1924	Scarlet fever	George F, DICK AND Gladys Dick
1924	Tetanus (tetanus toxoid, TT)	French Gaston Ramon, C. Zoeller and P. Descombey
1926	Pertussis (whooping cough)	Leila Denmark
1932	Yellow fever	Max Theiler and Jean Laigret
1937	Typhus	Rudolf Weigl, Ludwik Fleck and Zinsser
1937	Influenza	Anatol Smorodintsev
1941	Tick-borne encephalitis	
1952	Polio (Salk vaccine)	
1954	Japanese encephalitis	
1954	Anthrax	
1957	adenovirus-4 and 7	
1962	Oral polio vaccine (Sabin vaccine)	
1963	Measles	
1967	Mumps	
1970	Rubella	
1977	Pneumonia (Streptococcus pneumonia)	
1978	Meningitis (Neisseria meningitides)	
1980	Smallpox declared eradicated worldwide due to vaccination efforts	
1981	Hepatitis B (first vaccine to target a cause of cancer)	
1984	Chicken pox	
1985	*Haemophilus influenzae* type b (HiB)	
1989	Q fever	
1990	Hantavirus hemorrhagic fever with renal syndrome	
1991	Hepatitis A	
1998	Lyme disease	
1998	Rotavirus	
21st century		
2003	Influenza vaccine approved in US (Flu Mist)	
2006	Human papillomavirus (which is case of cervical cancer)	

Continued

Table 1.1 Showing few important developments in vaccination during 20th and 21st centuries—cont'd

Year of development	Type of diseases	Name of scientist
2012	Hepatitis E	
2012	quadrivalent (4-strain) Influenza vaccine	
2015	Enterovirus 71, one cause of hand foot mouth disease	
2015	Malaria	
2015	Dengue fever	
2019	Ebola approved	
2020	COVID-19	

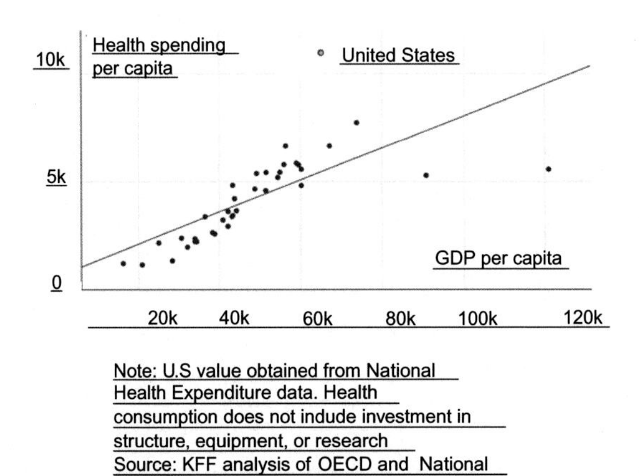

FIG. 1.11

GDP per capita and health consumption spending per capita, 2019 (US dollar, PPP adjusted).

On average, other wealthy countries spend about half as much per person on health than United States (Fig. 1.12).

China also ranks below the Organization for Economic Cooperation and Development (OECD) average in terms of health expenditure per capita, spending USD 480 in 2012, compared with an OECD average of USD 3484 (Table 1.2).

India 2020 population is estimated at 1,380,004,385 people at the midyear according to UN data. India population is equivalent to 17.7% of the total world population. Overall, India's public health expenditure has remained between 1.2% and 1.6% of GDP between 2008 and 09 and 2019–20.

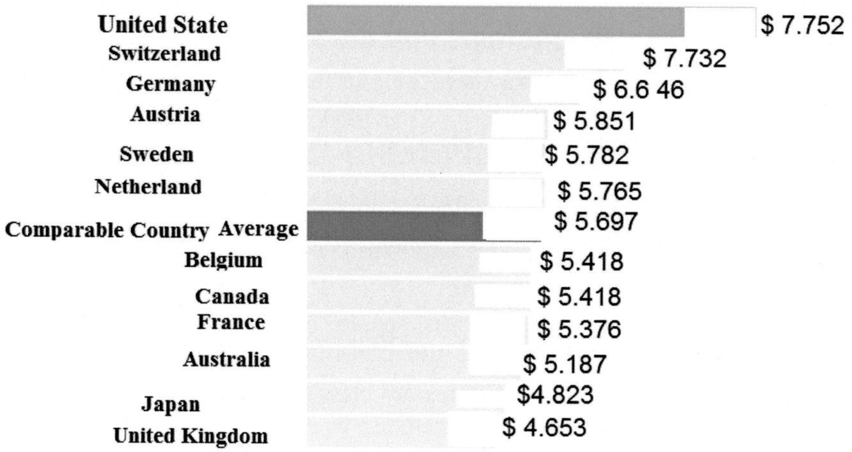

United State $ 7.752
Switzerland $ 7.732
Germany $ 6.6 46
Austria $ 5.851
Sweden $ 5.782
Netherland $ 5.765
Comparable Country Average $ 5.697
Belgium $ 5.418
Canada $ 5.418
France $ 5.376
Australia $ 5.187
Japan $4.823
United Kingdom $ 4.653

Note: U.S value obtained from National Health Expenditure data. Health consumption does not include investment in structure, equipment or research.. Source: KFF analysis of OECD and National Health Expenditure (NHE) data

FIG. 1.12

Showing health consumption expenditure per capita, US dollar PPP adjusted, 2019.

Table 1.2 Health expenditure per Chinese citizen from 2009 to 2019 (in Yuan) per capital health expenditure in yuan.

Year	Health expenditure (Yuan)
2019	4702.9
2018	4236.98
2017	3783.83
2016	3351.74
2015	290.8
2014	2582.60
2013	2327.37
2012	1806.95
2011	149.95
2010	1490.50
2009	1313.3

1.4 Types of health problems and diseases

Health problem refers to abnormal functioning of the health system due to some pathological infection or some other disorder without any pain. The synonyms of health problems may be ill health or unhealthiness. Overall condition of health can be categorized into four groups: (i) infectious diseases, (ii) deficiency diseases, (iii) hereditary diseases, and (iv) physiological diseases.

Diseases can also be classified as communicable versus noncommunicable diseases.

1.4.1 Diseases classification as per Britannica encyclopedia

As per Britannica Encyclopedia, diseases are classified as (Fig. 1.13): (1) topographic, by bodily region or system, (2) anatomic, by organ or tissue, (3) physiological, by function or effect, (4) pathological, by the nature of the disease process, (5) etiologic (causal), (6) juristic, by the speed of advent of death, (7) epidemiological, and (8) statistical. Any single disease may fall within several of these classifications.

(i) Topographic Classification

Under this category, diseases are subdivided into gastrointestinal diseases, vascular diseases, abnormal diseases, and chest diseases. Generally, doctors specialized in

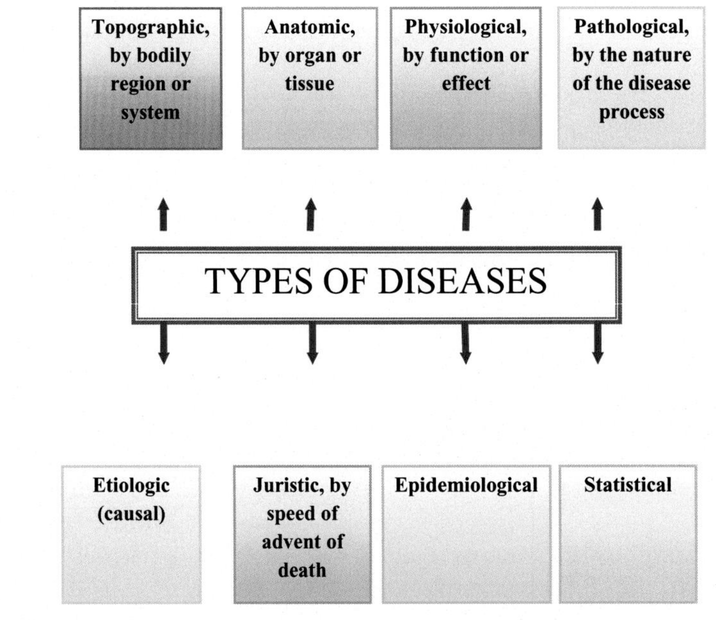

FIG. 1.13

Types of diseases as classified by Britannica Encyclopedia.

medicine, gastrointestinal diseases, chest diseases follow such topographic or systemic divisions.

(ii) Anatomical classification

This classification is mainly based on organs or tissue affected by diseases such as heart diseases, lung diseases, and liver diseases. Cardiologists, urologists, neurologists belong to this category. Even atherosclerosis of the coronary arteries come under this category of disease.

(iii) Physiological classification

This classification is mainly based on the functional disorder of any organs produced by a specific disorder. Examples include respiratory or metabolic diseases. Respiratory diseases are mainly involved in the interference of intake and expulsion of oxygen and carbon dioxide. Diabetes and gout problems are also under this category.

(iv) Pathological classification

The pathological classification is based on the effect of foreign bodies on the normal structure and function of the body as a whole or part. Neoplastic and inflammatory diseases are examples. Neoplastic disease includes the whole range of tumors, particularly cancers, and their effect on human beings.

(v) Etiologic classification of diseases

Etiology (from Greek word meaning the study of cause) in medicine is known as the determination of a cause of diseases or pathology. One disease entity can have more than one etiology that can lead to more than one disease. The etiology is mainly caused by biological interference. On this basis, diseases might be classified as an example, staphylococcal or rickettsial or fungal origin.

Etiology focuses on the back story of a disease. The illness caused by etiology can be due to intrinsic (coming from within), extrinsic (originating from external factors), or idiopathic (cause unknown). Etiology is not only disease specific but also person specific.

(vi) Juristic basis

The juristic basis of the classification of diseases is concerned with the legal circumstances in which sudden death occurs in which the cause of death is not known.

(vii) Epidemiological classification

Epidemiology is a Greek origin word: *epi*, meaning on or upon, *dermos*, meaning people, and *logos*, meaning the study of "*Epidemiology is the study of the distribution and determinant of health-related states or events in Specific population, and the application of this study to the control of health problem*" [19].

Originally, epidemiology focused exclusively on epidemics of communicable diseases [2] but was subsequently expanded to address endemic communicable diseases and noncommunicable diseases.

(viii) Statistical disease classification

The statistical basis of classification of disease is mainly based on analysis of the incidence of a particular disease and occurrence rate after a specific time period. For example, if a disease occurrence is 100 in number, and a period of 3 years, the prevalence would be 300. By statistical analysis, it would be possible to understand the possible cause of disease on the basis of cumulative data analysis related to food habits, or epidemiological, nutritional, and pathological analysis. The statistical analysis is on the high level of fats and carbohydrates may be the cause of atherosclerosis.

1.4.2 Diseases classification on contamination basis

Communicable diseases

Disease is an abnormal condition of health that may be due to interference of external factors such as (i) pathogens or (ii) internal dysfunctions. Infectious diseases are transmitted from a person by direct or indirect means. Some of the important contagious diseases include: COVID-19 (2019-nCoV in which "n" is for novel and "CoV is for coronavirus"), Norovirus (Stomach Flu), Influenza, Meningitis, pertussis, sexually transmitted diseases, methicillin-resistant *Staphylococcus aureus* (MRSA).

Decline in communicable diseases

As stated in WHO 2019 report, there was a remarkable decline in deadly diseases such as pneumonia and other lower respiratory infections which were the fourth leading cause of death in 2000. The global number of death due to such diseases has decreased by nearly half a million [19].

In 2019, globally, HIV/AIDS dropped from the 8th leading cause of death as compared with the death rate in 2000. This could have been possible due to the success of efforts to infection, test for the virus, and treatment of the diseases over the last two decades. About 6.2% of the world's population is contaminated with HIV. In 2018, there were 800,000 new HIV infections, worldwide [20].

East and Southern Africa is the region hardest hit by HIV. In 2018, South Africa accounted for more than a quarter (240,000). In addition, seven other countries accounted for more than 50% of new infections: Mozambique (150,000), Tanzania (72,000), Uganda (53,000), Zambia (48,000), Kenya (46,000), Malawi (38,000), and Zimbabwe (38,000) [20]. Overall, new infections in the region have declined by 28% since 2010 [21]. In this region, in 2018, the number of deaths has fallen by 44% since 2014 [21]. In 2018, 85% of people living with HIV were aware of their status, and 79% of them were on treatment (equivalent to 67% of all people living with HIV in the region) [22].

In 2019, the global death rate of tuberculosis was not in the top ten death rate, as compared to its 7th place in 2000. There was about a 30% reduction in global death. But, it still remains among 10 causes of deaths in the African and South-East Asian regions, where it is the eighth and fifth leading cause, respectively.

WHO reports also highlight that the total communicable diseases still persist in low-income countries. Six of the top 10 causes of death in low-income countries are still communicable diseases, including malaria (6th), tuberculosis (8th), and HIV/AIDS (9th). But, at the global level, the overall decrease in the death rate due to communicable diseases has gone down.

Current scenario in COVID-19

It was in December 2019, the first report on coronavirus-related pneumonia (SARS-Co-2) appeared in the public domain and created havoc. It was on January 11, 2020, the first death was recorded in Wuhan, China, and by this time global deaths were set to reach 2 million. Within a period of 4 months the death rate reached 1 million on September 28, 2020, to 2 million on January 15, 2021; people in over 210 countries were conformed with COVID-19 and 2 million people died.

List of top 10 communicable diseases

Following are the top 10 communicable diseases, as reported by WHO: 2019-n CoV, CRE, Ebola, Enterovirus D68, Flu, Hantavirus, Hepatitis A, Hepatitis B, HIV/AIDS, Measles, MRSA, Pertussis, Rabies, Sexually Transmitted Diseases, Shigellosis, Tuberculosis, West Nile Virus, and Zika.

Noncommunicable diseases

Mainly, noncommunicable diseases are responsible for the cause of more death as compared to communicable diseases. As reported by WHO [23], in 2019, the top 10 causes of death accounted for about 55% of the 55.4 million deaths worldwide. Mainly, the three broad groups of diseases associated with the 10 top global death are (i) cardiovascular (ischemic heart diseases, stroke), (ii) respiratory (chronic obstructive pulmonary disease, lower respiratory infections), and (iii) neonatal conditions which include birth asphyxia and birth trauma, neonatal sepsis and infection, and preterm birth complications.

The last two decades have been witnessing heart disease as the leading cause of death at the global level. The number of deaths from heart disease increased by more than 2 million since 2000, to about 9 million in 2019. Heart is reported as 16% of the total deaths from all causes. The WHO Western Pacific Region is intensely under the grief of noncommunicable diseases (NCDs) like cancer, diabetes, and chronic respiratory diseases. But European region has seen a relative decline in heart diseases, with death falling by 15% (www.who.int/countries for the list of countries in each WHO region).

Alzheimer's and other forms of dementia are on the list of the top 10 globally acclaimed NCDs. The death due to Alzheimer has occupied third in both the Americas and Europe in 2019. The female death rate due to this disease is about 65% as compared to the male death rate. Death due to diabetes has increased by 70% globally between 2000 and 2019. In the Eastern Mediterranean, deaths from diabetes have more than doubled and represent the greatest percentage increase of all WHO regions.

WHO presents the most comprehensive and update information on population health, including life expectancy, healthy life expectancy, mortality and morbidity, status of diseases, at the global level (Fig. 1.14). Information on these issues results

FIG. 1.14

Concept on population health management.

from data survey of WHO from the best available resources around the world. Robust health data are critical to have control and prevention of diseases at the family, community, and global level. In addition, a global health data bank is critical for analyzing the health impact on the economic status of a nation. As of today, COVID-19 has tragically claimed more than 1.5 million lives. People living with preexisting health conditions such as heart diseases, diabetes, and respiratory diseases are at higher risk of complications due to COVID-19.

1.4.3 The international classification of diseases

The International Classification of Diseases (ICD) is a global body maintained by the World Health Organization (WHO), directed and coordinated by the authority of health within the United Nations System. Presently, ICD is a globally used diagnostic tool for epidemiology, health management, and clinical purposes. The ICD is a core statistical-based diagnostic system for health-care-related issues of the WHO Family of International Classification (WHO-FIC).

Conceptual development of ICD

In 1860, during the international statistical congress, Florence Nightingale proposed a developing model on international health data analysis. Subsequently, in 1893, Jacques Bertillon, a French Scientist, at a congress held in Chicago, proposed a comprehensive model on causes of death entitled "Bertillon Classification of Causes of Death." A number of countries adopted Dr. Bertillon's system. In 1898, the American Public Health Association (APHA) recommended Canada, Mexico, and United States to also adopt. Besides this, APHA also recommended revising the system every 10 years in order to update with current development in this regard. In August

1900, for the first time, a conference on Revision of the Bertillon or International List of causes of death was held in Parish. Delegates from 26 countries participated in this conference. In August 1900, a detailed classification of causes of death with 179 groups, and a shorten classification of 35 groups were framed. In subsequent years, series of meetings starting from 1910 were held, with a gap of 10 years (Table 1. DDD). The French Government called succeeding conferences in 1920, 1929, and 1938 to revise the classification for health-related death for six times and resulted in the development of two volumes. The final getup of these two volumes contained mortality and mortality conditions with the title "Manual of International Statistical Classification of Diseases, Injuries and Causes of Death (ICD)."

The six revisions were related to morbidity and mortality conditions, and the title was changed, accordingly: "Manual of International statistical classification of Diseases, Injuries and Causes of Death (ICD)." In 1948, the WHO had taken responsibility for developing and publishing revisions to the ICD. At a later stage, WHO had also taken the responsibility of revising and publishing seventh and eighth in 1957 and 1968, respectively. Subsequently, the ninth revision of the ICD (ICD-9) was published in 1978. Later on, the US Public Health Service made a modification in ICD-9 in order to meet the needs of American hospitals and called it International Classification of Diseases, Ninth Revision, Clinical Modification (ICD-9-CM).

In May 1990, the 43 World Health Assembly recommended ICD-10. It is cited in more than 20,000 scientific articles and used by more than 150 countries around the world. It was translated into more than 40 languages.

The ICD-10 consists of

- Detail tabular list containing cause-of-death, and code number (Volume 1)
- Inclusion and exclusion terms for cause-of-death title (Volume 1)
- An alphabetical index to diseases and nature of the injury, external causes of injury, table of drugs and chemicals (Volume 3)
- Description, guidelines, and coding rules (Volume 2).

The conversion of ICD-9 into ICD-10 has changed the presentation of mortality data systems which affect the instruction manuals, medical software, and analysis.

ICD-11 update

ICD-11 has been adopted by the 72 World Health Assemblies in May 2019 and will come into effect on January 1, 2022. ICD-11 introduced two striking features: extensions and clustering which would be helpful in post-cording and the addition of specific detail to coded entities. Both features have the potential to improve ICD-11 code data. The ICD-11 catalog is most important to specify human diseases, medical conditions, and mental health disorders and is used for insurance coding purposes, for statistical tracking of illnesses, and as a global health categorization tool that can be used across countries and in different languages.

The WHO keeps update responsibility in modifying or further developing ICD-11, on a need basis. It is due to the involvement of 300 specialists from 55 countries divided into 30 work groups.

Who can use ICD-11

Mostly, doctors, nurses, research scientists, health professors, workers involved in health-care management and technology, policy makers, and health insurance companies use ICD-11. All Member States use ICD. Presently, ICD is available in 43 languages. It is also applicable to systematically present mortality data as a primary indicator of health status. All Member States are entitled to use the most updated information on ICD from the public domain.

Important of ICD-11

The most significant achievement of ICD is its availability in different languages in order to attract global attention. This is immensely helpful having a comparative data structure for sharing, at a global level. ICD is the most reliable and evidence-based data bank in making a critical decision. The revision of ICD-11 for time and again is immensely helpful to keep updated progress in health sciences and medical practice. Through the application of the latest electronic communication, ICD-11 can be applicable in more efficient ways with minimum time. ICD-11 also works well for collaborative web-based editing that is open to all interested parties.

1.5 Outbreaks, epidemic, and pandemics

In epidemiology terminology, a sudden increase in the occurrence of a specific disease for a particular time and place is known as an outbreak. Outbreaks include epidemic, which has restricted use for infectious diseases or environmentally origin diseases like water or foodborne disease which may affect a locality in a country or a group of countries. Pandemics are near-global disease outbreaks when multiple countries across the world are infected (Fig. 1.15).

1.5.1 Outbreaks

As stated above, outbreaks may last for few days or linger for months or several years (Fig. 1.16). Some outbreaks are supposed to be yearly basis such as influenza. Sometimes a single case of an infectious disease may be considered outbreak (e.g., food-borne botulism or bioterrorism agent such as anthrax).

On February 12, an outbreak of the Ebola virus in the Democratic Republic of the Congo was a great concern for Public health emergency of international concern (PHEIC). The Committee acknowledged the outbreaks as high at a national and regional level and low at the global level.

Causes of diseases outbreaks

Generally, the most common causes of disease outbreaks include infection, transmission through person-to-person contact, animal-to-personal contact, or from the environment or other media. Expose to chemical or radioactive materials may also cause outbreaks of disease. Communicable diseases can also be transferred

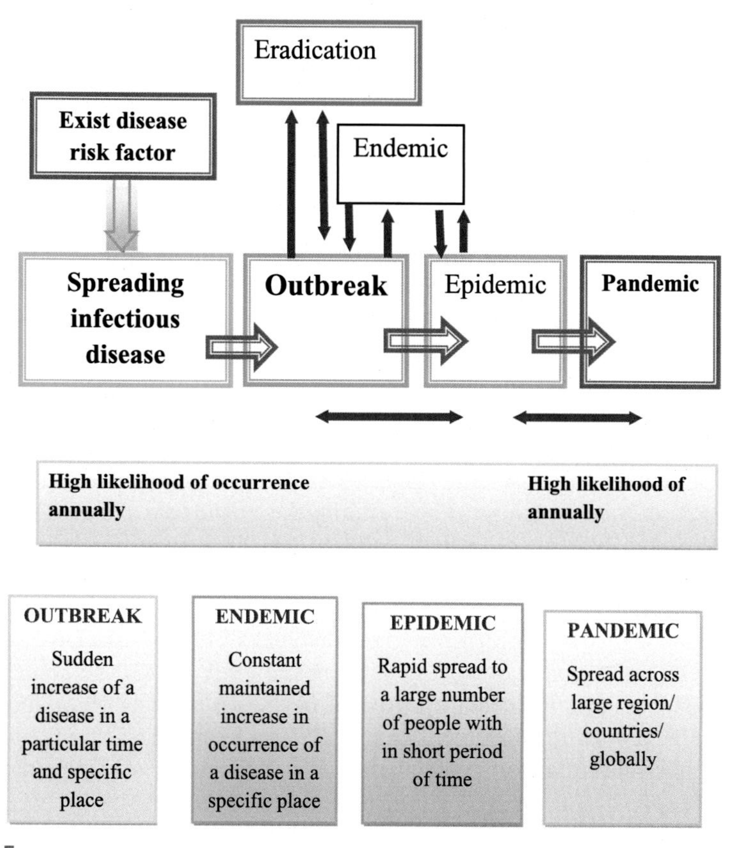

FIG. 1.15

Pandemic and spread infection disease.

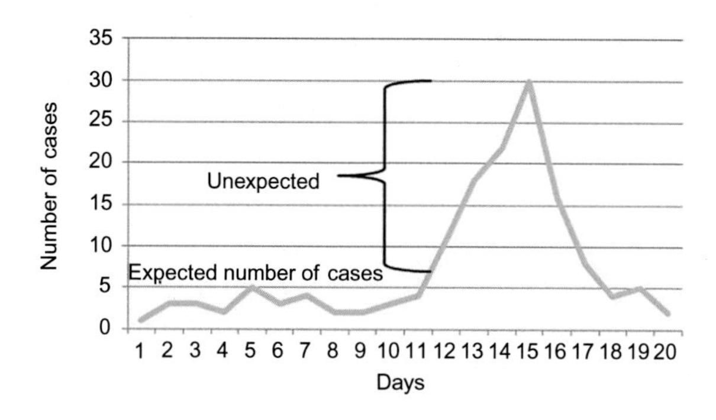

FIG. 1.16

Showing outbreak of a disease in which occurrence of more cases of a disease than expected for a particular place and time prior to taking any preventive measures.

Source: Centre for Disease Control and Prevention (CDCP), United State. Disease clusters: an overview case definition from the United States Department of Health and Human Services http://www.atsdr.cdc.gov/HEC/CSEM/cluster/case_def.html.

through water, sanitation, and food and air quality. Poor hygiene, poor living conditions, mismanagement in waste disposal facilities, can also cause diarrheal diseases. These diseases are a major cause of suffering and death in an emergency situation.

1.5.2 Epidemic

Epidemic is referred to as the process of rapid spreading of an infectious disease to a large number of people in a given population within a short period of time. The Centers for Disease Control and Prevention (CDCP) defines epidemic as: "an unexpected increase in the number of disease cases in a specific geographical area." So, an epidemic is any rise in cases beyond the baseline for that geographic area.

In 2003, the severe acute respiratory syndrome (SARS) epidemic took the lives of nearly 800 people worldwide. The Plague of Athens was an epidemic that devastated the city-state of Athens in ancient Greece during the second year of the Peloponnesian War. The plague killed an estimated 75,000 to 100,000 people, around one-quarter of the population, and is believed to have entered Athens through Piraeus, the city's port and sole source of food and supplies.

On March 23, 2014, WHO declared the outbreak of Ebola Virus Disease (EVD) in the forest rural region of Southeastern Guinea. By July 2014, the outbreak spread to the capital of all three countries. A total of 28,616 cases of EVD and 11,310 deaths were reported in Guinea, Liberia, and Sierra Leone. This was the first time EVD extended out from more isolated, rural areas and into densely populated urban centers.

There are four types of influenza viruses: A, B, C, and D. Human influenza A and B viruses cause seasonal epidemics of disease almost every winter in the United States. Influenza A viruses are the only influenza viruses known to cause flu pandemics, i.e., a global epidemic of flu disease.

Based on the combination of surface proteins hemagglutinin (HA) and the neuraminidase (NA), influenza A is classified into subtypes: A(H1N1) and A (H3N2) influenza viruses. The A1(H1N1) is also written as A(H1N1)pdm09 as it caused the pandemic in 2009. Only influenza type A viruses are known to have caused the pandemic. But Influenza B viruses are without any subtypes. Influenza C virus is rarely seen and has also no important in public health. Influenza D is restricted to cattle and not infectious to humans.

Symptoms

Sudden onset of fever, dry cough, headache, muscle, and joint pain, severe malaise (feeling unwell), sore throat, and runny nose are some of the important symptoms of seasonal influenza. Illnesses range from mild to severe and even death. Generally, people from high-risk groups are hospitalized or die. Worldwide, these annual epidemics are estimated to result in about 3–5 million cases of severe illness, and about 290,000–650,000 respiratory deaths. In industrialized countries most deaths associated with influenza occur among people age 65 or older [24,25].

Causes of epidemic

Generally, epidemics of infectious diseases are caused by various factors like change in the ecology of the host population, a genetic change in the pathogen reservoir, or the introduction of an emerging pathogen to a host population. Due to sudden reduction in immunity potential of the host population caused by a new pathogen or genetically mutated pathogen may be responsible for epidemic outbreaks in a particular locality.

Infected food supplies such as contaminated drinking water and the migration of populations of certain animals, such as rats or mosquitoes (disease vectors) are also cause of the epidemic.

Generally, the epidemic outbreak is seasonal. Due to seasonal change, the physiology of the body and its immune system gets change. In temperature climates, a seasonal epidemic occurs mainly during winter, while in tropical regions, influenza may occur throughout the year, causing outbreaks more irregularly. The time from infection to illness, known as the incubation period, is about 2 days but ranges from 1 to 4 days.

Smallpox, cholera, yellow fever, typhoid, measles, and polio are some of the worst epidemics in American history.

Types of epidemics

Based on propagation, an epidemic can be classified into three groups: (i) common-source outbreaks, (ii) transmission (propagated or progressive epidemic), and (iii) mixed epidemics.

(i) Common-source outbreaks

Common-source epidemics are mainly due to exposure to an infectious agent, and occur frequently but not always. They are generated from the contaminated environment (air, water, food, soil) by industrial pollutants. For example, the Bhopal gas tragedy in India and Minamata disease in Japan resulting from the consumption of fish containing a high concentration of methyl mercury.

(a) *Single exposure or "point-source" epidemic*

If the exposure is singular, and all of the affected individuals develop the diseases over a single exposure, it can be termed a "point-source outbreak" (epidemic). Point-source outbreaks tend to have an epidemic curve with a rapid increase in cases followed by a somewhat slower decline, and all of the cases tend to fall within one incubation period. The graph below from a hepatitis outbreak is an example of a point-source epidemic (Fig. 1.17).

The incubation period for hepatitis ranges from 15 to 50 days, with an average of about 28–30 days. In point-source epidemic of Hepatitis, the rise and fall of new cases occur within about 30 days span of time.

The exposure to the disease agent is brief and essentially simultaneous; the resultant cases all develop within one incubation period of the disease (an epidemic of food poising). In a "point-source" the epidemic curve rises and falls rapidly without any secondary waves. In addition, the epidemic tends to be explosive, there is cluster of cases within a narrow interval of time, more importantly, all the cases develop within one incubation period of the diseases.

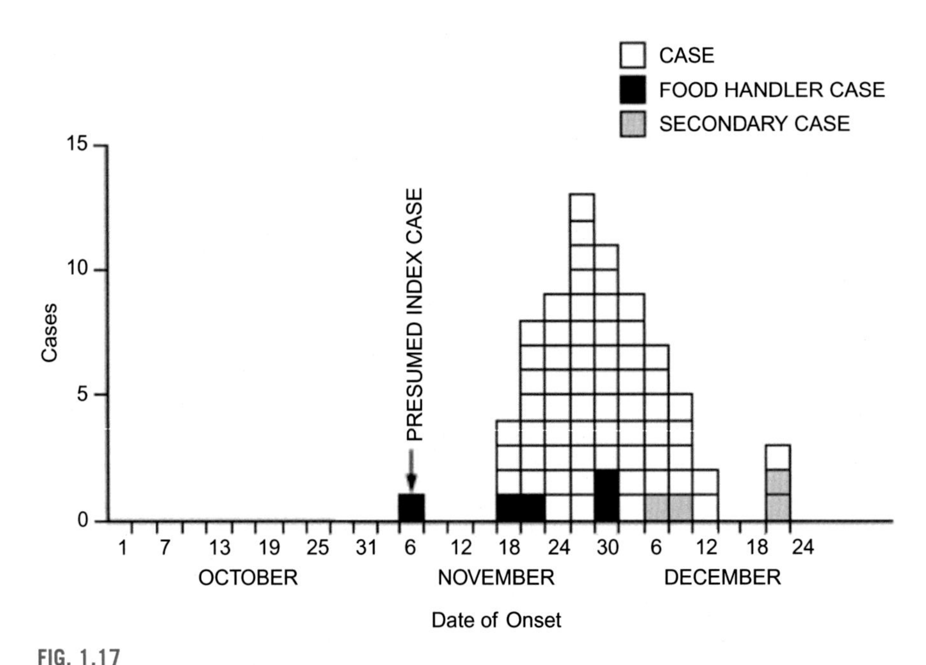

FIG. 1.17

Showing mode of spreading epidemic "point source."

(b) *Continuous or multiple exposure epidemics*

If the exposure is continuous or variable, it can be termed a continuous outbreak or intermittent outbreak, respectively (Fig. 1.18).

Sometimes the exposure from the same source may be prolonged, and continuous, repeated, or intermittent. For example, a prostitute may be a common source in a gonorrhea outbreak, but since she infects her client over a period of time, there may be no explosive rise in the number of cases. A single source of contaminated water, or a nationally distributed brand of vaccine (polio vaccine), or food may be responsible for similar outbreaks.

The spreading of disease from person to person is known as a propagated outbreak. Affected individuals act as independent reservoirs, as with syphilis. Transmission may also be vehicle borne (e.g., transmission of hepatitis B or HIV by sharing needles) or vector borne (e.g., transmission of yellow fever by mosquitoes).

If the exposure is singular, and all of the affected individuals develop the diseases over a single exposure, it can be termed a point-source outbreak. If the exposure is continuous or variable, it can be termed a continuous outbreak or intermittent outbreak, respectively.

(ii) Transmission (propagated or progressive epidemic)

Transmission is a process of propagation of disease from person to person, directly or indirectly (Fig. 1.19). Direct contact transmission occurs when there is

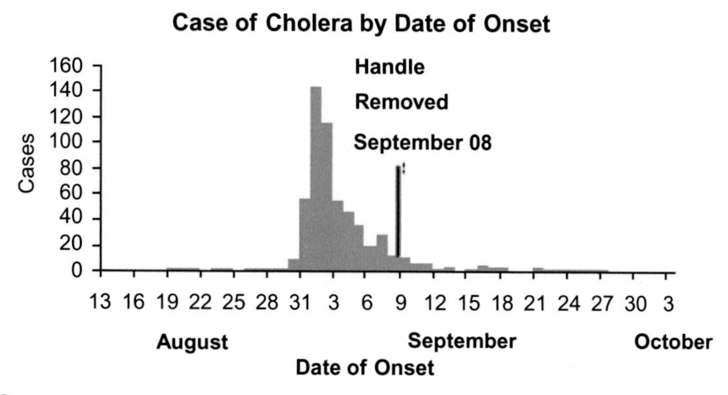

FIG. 1.18

The epidemic curve shows the cholera outbreak in the Broad Street area of London in 1854. Cholera has an incubation period of 1–3 days, and even though residents began to flee when the outbreak erupted, which can be seen that this outbreak lasted for more than a single incubation period.

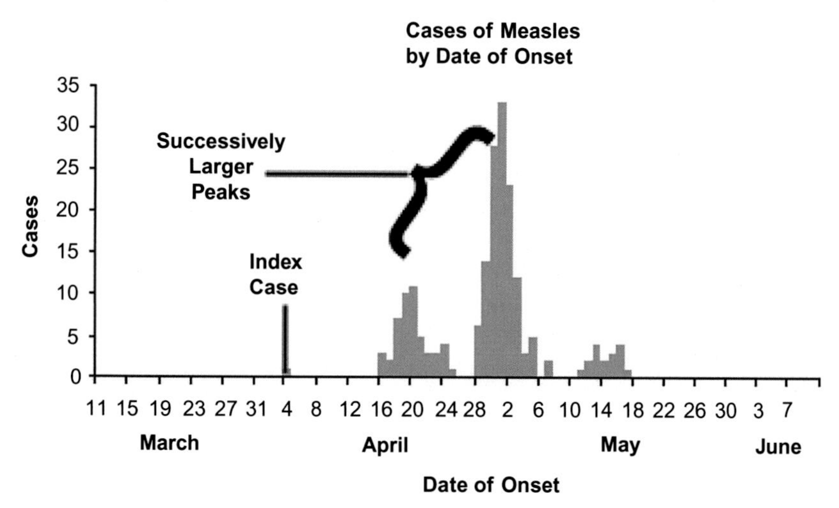

FIG. 1.19

Showing the epidemic curve from an outbreak of measles that began with a single index case who infected a number of other individuals. (The incubation period for measles averages 10 days with a range of 7–18 days).

physical contact between an infected person and a susceptible person. Indirect contact transmission occurs when there is no direct human-to-human contact. Contact occurs from a reservoir to contaminated surfaces or objectives, or to vectors such as mosquitoes, flies, mites, fleas, ticks, rodents, or dogs.

(a) *Spreading of infection*
Direct contact infections spread when pathogens pass from the infected person to the healthy person via direct physical contact with blood or body fluids. Examples of direct contact are touching, kissing, sexual contact, contact with oral secretions, or contact with body lesions. Indirect contact infections spread when an infected person sneezes or coughs, sending infectious droplets into the air. Droplets generally travel between 3 and 6 ft and land on nearby surfaces or objects including tables, doorknobs with their hands, and then touch their eyes, nose, or mouth.
A lot of diseases spread through contact transmission. Examples are chicken pox, common cold, conjunctive (pink eyes), hepatitis A and B, herpes, adeno/rhinovirus, neisserial, meningitides, and mycoplasma pneumonia.

(b) *The means of transmission*
Contact (direct and/or indirect), droplet, airborne, vector, and common vehicle are some of the important means of disease transmission processes. The portal of entry is the means by which the pathogens gain access to the new host. This can occur, for example, through ingestion, breathing, or skin puncture (Fig. 1.20A and B).

1.5.2.1 How to avoid

Earlier (Section 1.2.3), it has already been mentioned how frequent hand washing, use of mouth mask while out of the home. It is also recommended for regular disinfection of frequently touched surfaces such as doorknobs, handles, handrails, phones, office supplies, and children's toys. Using barriers such as gloves, masks, or condoms can help avoid the spread of germs.

(iii) Mixed epidemic

A mixed epidemic includes both common-source and propagated outbreak characteristics. For example, people infected through a common-source outbreak might later transmit or spread the disease through direct contact with others. Mixed epidemics are often caused by food-borne infectious agents.

1.5.3 Pandemic

A pandemic is an epidemic that spreads over multiple countries or continents. The world population have been confronting with a number of devastating pandemics diseases such as smallpox, tuberculosis, plague, etc. (Table 1.3).

Coronavirus

Coronaviruses belong to a large family of RNA viruses that have been around for a long time. Many of them can cause a variety of illnesses, from mild cough to severe respiratory illnesses. In humans and birds, they cause respiratory tract infections that can range from mild to lethal. Mild illnesses in humans show common cold with mild fever. The lethal variety of corona can cause SARS (Severe Acute Respiratory Syndrome), MERS (Middle East Respiratory Syndrome), and COVID-19.

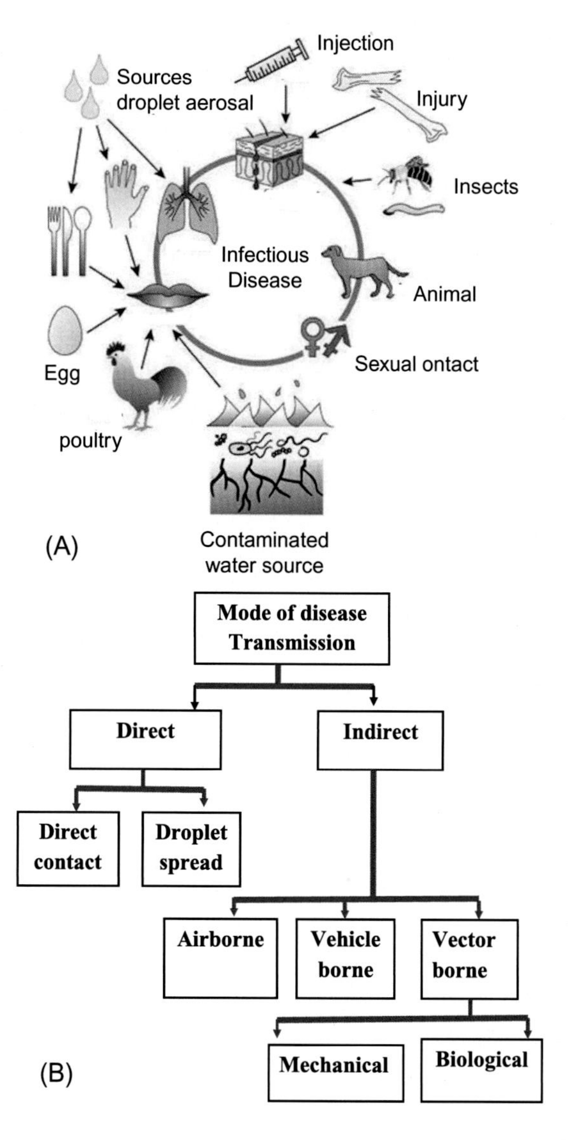

FIG. 1.20

(A, B) Showing mode of transmission of diseases.

The novel coronavirus causing COVID-19 is the one known to infect humans severely. Earlier, it is probably been around for some time in animals, and in due time crossed over into people. So, this virus is not new to the world, but it is new to humans. In 2019, it was known as novel coronavirus (n COV), which was subsequently named the "COVID-19" ("CO" stand for corona, "VI" for virus, and "D" for diseases) or SARS-CoV-2.

Table 1.3 Major epidemics and pandemics death tool.

Rank	Epidemics/pandemics	Death toll	Location
1	Black death 1346–1353	75–200 million	Europe, Asia, and North Africa
2	Spanish flu 1918–1920	17–100 million	Worldwide
3	Plague of Justinian 541–549	15–100 million	Europe and West Asia
4	HIV/AIDS pandemic 1981–present	35 million+ (as of 2020)	Worldwide
5	Third plague pandemic 1855–1960	12–15 million	Worldwide
6	Cocoliztli epidemic 1545–1548	5–15 million	Mexico
7	Antonine plague 164–180	5–10 million	Roman Empire
8	Mexico smallpox epidemic 1519–1520	5–8 million	Mexico
9	COVID-19 pandemic 2019–present	2.93 million+ (as of April 2021)	Worldwide
10	Russia typhus epidemic 1918–1922	2–3 million	Russia
11	Influenza pandemic 1957–1958	1–4 million	Worldwide
	Hong-Kong flu 1968–1969	1–4 million	Worldwide
13	Cocoliztli epidemic 1576–1580	2–2.5 million	Mexico
14	Japanese smallpox epidemic 735–737	2 million	Japan
	Persian plague 1772–1773	2 million	Persia
16	Naples plague 1656–1658	1.25 million	Southern Italy
17	Third cholera pandemic 1846–1860	1 million+	Worldwide
18	Italian plague 1629–1631	1 million	Italy
	1889–1890 flu pandemic 1869–1890	1 million	

COVID-19 vs influenza

Both the coronavirus and influenza cause respiratory disease and transmit by droplets, and fomites. Yet there are important differences between the two viruses and how they spread (Table 1.4).

Table 1.4 Differences between corona virus and influenza.

Corona virus	Influenza
The serial interval for COVID-19 virus is estimated to be 5–6 days	Influenza has a shorter median incubation period (the time from infection to appearance of symptoms) and a shorter serial interval (the time between successive cases) Influenza virus, the serial interval is 3 days
There are cases who can shed COVID-19 virus 24–48 h prior to symptom onset	Transmission of the virus before the appearance of symptoms is a major driver of transmission for influenza
The number of secondary infections generated from one infected individual is understood to be between 2 and 2.5 for COVID-19 virus	In case of influenza the secondary infections generated from one infected individual is less than that of corona virus
COVID-19 virus, initial data indicates that children are less affected than adults and that clinical attack rates in the 0–19 age group are low. Further preliminary data from household transmission studies in China suggest that children are infected from adults, rather than vice versa	Children are important drivers of influenza virus transmission in the community
For COVID-19, data to date suggest that 80% of infections are mild or asymptomatic, 15% are severe infection, requiring oxygen and 5% are critical infections, requiring ventilation	Fractions of severe and critical infection is lesser in influenza
Older age and underlying conditions increase the risk for severe infection	Risk for severe influenza infection are children, pregnant women, elderly, those with underlying chronic medical conditions

COVID-19

Outbreak

Retrospective investigations by Chinese authorities have identified human cases with the onset of symptoms in early December 2019. While some of the earliest known cases had a link to a wholesale food market in Wuhan.

On 30 January, WHO declared the COVID-19 outbreak a public health emergency of international concern (PHEIC), as the highest level of alarm. At that time, there were 98 cases and no deaths in 18 countries outside China.

On 11 March, WHO declared COVID-19 as a pandemic due to its rapid spread all around the world. By then, more than 11,800 cases and 4292 deaths in 114 countries had been reported. Surprisingly, by mid-March 2020, WHO European Region had become the epicenter of the epidemic, reporting over 40% of globally confirmed cases. As of April 28, 2020, 63% of global mortality from the virus was from the region. By April 2021, the total cases were reported about 138, 199,138; total deaths 2,975,571; and total recovered 111,170,872.

Structure

Genotypically, animal and plant viruses are two types. Herpes, wart viruses, and adenovirus contain long DNA molecules, whereas coronavirus has RNA as genetic material. Human coronaviruses' particles are spherical, 120–160 mm diameter, are named for their "sun-like" shape observed in the electron microscope. Influenza virus, HIV, rhinoviruses (common cold), SARS-CoV-2 (COVID-19) also contain RNA as genetic material. Onwards, we will be giving emphasis and restricted our discussion on the structure of CORONA-19, and how it attacks host cells.

The COVID-19 consists of RNA polymers tightly enveloped with protective protein molecules known as capsid proteins. In coronavirus, these proteins are called nucleocapsid (N). The core particle of coronavirus is further surrounded by an outer membrane envelope made of lipid (fat) with proteins inserted. This membrane is derived from the cells in which the virus was last assembled, and modified to contain specific viral proteins, including the spike (S), membrane (M), and envelope (E) proteins. A specific set of the proteins are projected on the outer surface of the particle and are known as spike proteins (S). The spike (S) protein has two subunits, e.g., the S1 subunit and S2 subunit. The S1 subunit has a receptor-binding domain that binds with the host cell receptor containing angiotensin-converting enzyme 2, and the S2 subunit mediates fusion between the viral and host cell membranes (Fig. 1.21).

Receptor binding

The S1 subunit has a receptor-binding domain that binds with the host cell receptor containing angiotensin-converting enzyme 2, and the S2 subunit mediates fusion between the viral and host cell membranes. Based on genome sequencing and pair wise protein sequence analysis, 2019-nCoV is recognized as SARC-related coronaviruses [26].

The viral genes can then enter the host cell to be copied, producing more viruses. Alike the SARS-2002, SARS-CoV-2 spikes bind to receptors on the human cell surface called angiotensin-converting enzyme 2 (ACE2). The SARS-CoV-2 spike was 10–20 times more likely to bind ACE2 on human cells than the spike from the SARS virus from 2002. This may enable SARS-CoV-2 to spread more easily from person to person than the earlier virus.

The SARS-CoV-2 (COVID-19) diffuses by respiratory droplets, as it was already demonstrated for other pathogens such as SARS-CoV, Middle East respiratory syndrome coronavirus (MERS), and influenza viruses [27–29]. The SARS-CoV-2 spike glycoprotein binds to ACE-2 and forms a potential target for developing specific drugs, antibiotics, and vaccines. It is also helpful in keeping a balance between RAS and ACE2/MAS [26,30–35]. Both 2019-nCoV and SARS-CoV enter the host cells via the same receptor, angiotensin-converting enzyme (ACE_2). Therefore, this virus was subsequently renamed SARS-CoV-2.

ACE2 mediates SARS-CoV-2 (COVID-19). COVID-19 enters the human body mainly through the SCE2+ TMPRSS2+ nasal epithelial cells (Fig. 1.22). The nasopharynx-associated lymphoid tissues (NALT) system is mainly responsible to first recognize exogenous airborne agents.

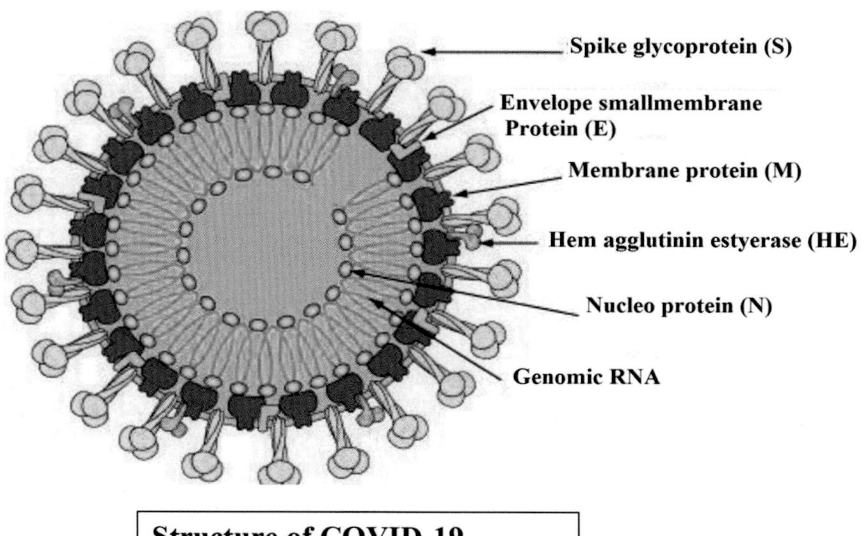

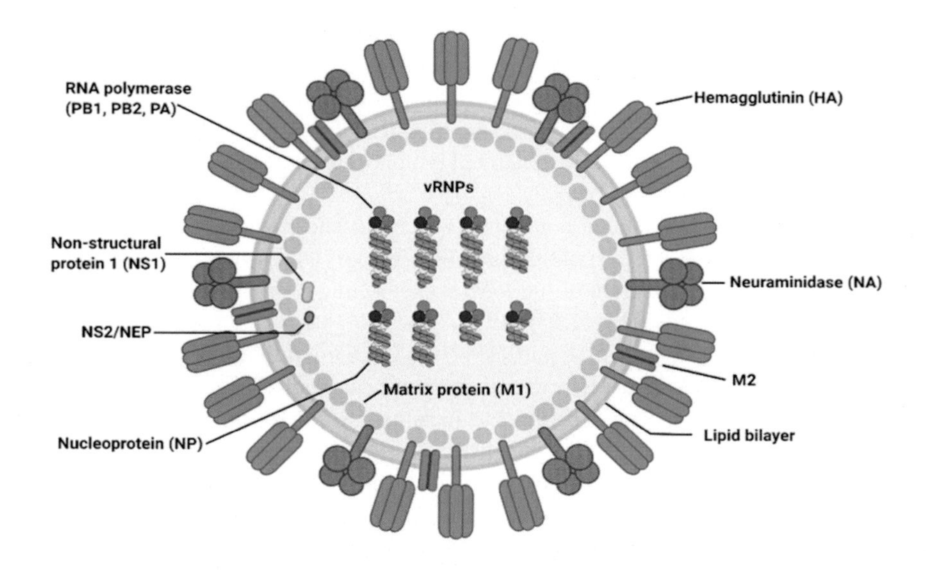

FIG. 1.21

Structure of COVID-19 virus as compared to the structure of influenza virus.

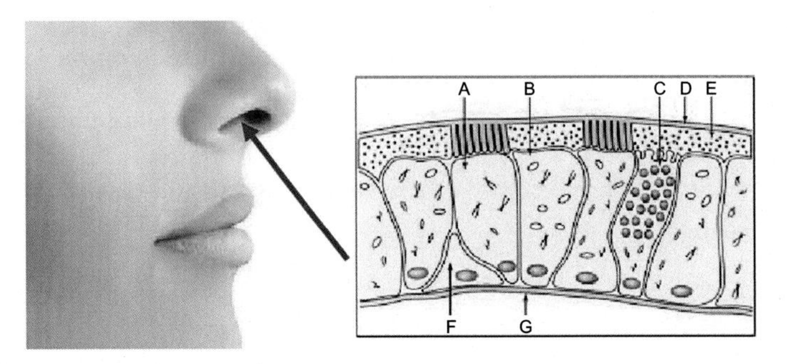

FIG. 1.22

Entry of epithelial cell of nose for CORONA virus.

The olfactory epithelium, found within the nasal cavity, contains olfactory receptor cells, which have specialized cilia extensions. The cilia trap odor molecules as they pass across the epithelial surface.

The NALT represents an immune system of the nasal mucosa and is a part of mucosa-associated lymphoid tissue (MALT) in mammals. It protects the body from airborne viruses and other infectious agents.

The nose represents an important component of the mucosal immunity in upper airways (UA) and is responsible for host protection and immune homeostasis between the commensal microbiota and invading pathogens (Fig. 1.22). The nose and NALT play a central role in the induction of mucosal immune responses, including the generation of Th1- and Th2- polarized lymphocytes and IgA-committed B cells [36–38].

Besides this, other cellular components such as dendritic cells (DCs), microfold (M) cells, and macrophages, nasal epithelial ciliated, and goblet cells are also associated within the induction of the local and systemic response to a wide range of pathogens and allergens (Fig. 1.23) [39].

In infection, the coronavirus particle serves three important functions for the genome: first, it provides the means to deliver the viral genome across the plasma membrane of a host cell; second, it serves as a means of escape for the newly synthesized genome; third, the viral particle functions as a durable vessel which protects the genome integrity on its journey between cells [40].

The infection of coronavirus with special reference to COVID-19 performs three important functions for genomes: (i) it provides the means to deliver the viral genome across the plasma membrane of a host cell, (ii) it serves as a means of escape for newly synthesized genome, and (iii) the viral particles function as a durable vessel which protects the genome integrity on its journey between cells [40].

The novel coronavirus SARS-CoV-2 (COVID-19) attacks the human body mainly through ACE2 + TMPRSS2 + nasal epithelial cells. The NALT acts as an immune system to first recognize exogenous airborne agents. In the human body, it is located in the most cranial pharyngeal mucosa. In addition, being in direct physical contact with the external environment, and rudely filters, moistens, and warms the inhaled

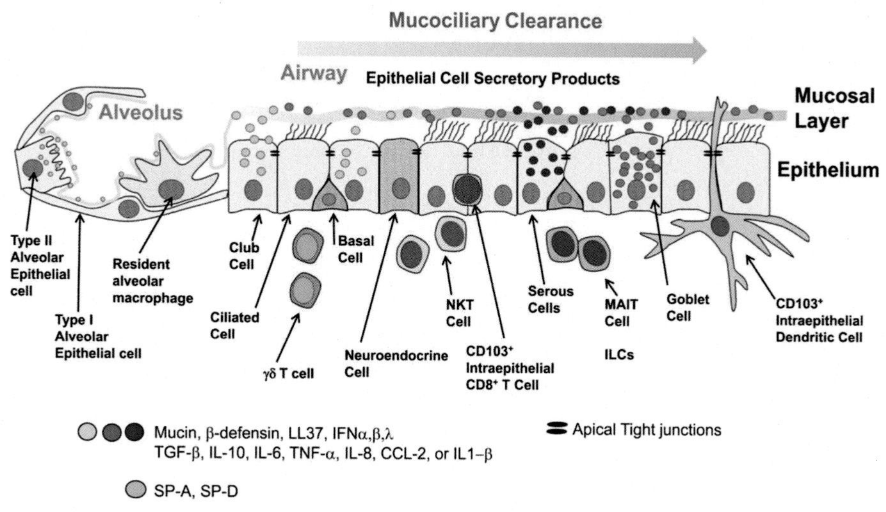

FIG. 1.23

The respiratory epithelium, and relevant epithelial-associated immune cells and soluble factors involved in IAV infection.

With courtesy from Takeda, K. et al. Allergic conversion of protective mucosal immunity against nasal bacteria in patients with chronic rhinosinusitis with nasal polyposis. J Allergy Clin Immunol 2019;143:1163–1175.

air to minimize the irritative effects on lower airways, to maintain the mucociliary clearance and to favor gaseous exchanges.

Following are few important steps on how the coronavirus (COVID-19) attack the human host cells, and subsequently complete its life cycle inside the host cells, before emerging out from the infectious cells.

Step 1

The viral spike (S) binds with ACE2 protein present on the surface membrane of epithelial cells and leads the virion into the host cells. ACE2 is an angiotensin-converting enzyme 2 that acts as a receptor for coronaviruses, including COVIS-19. ACE2 is commonly known as a cellular doorway for coronaviruses (Fig. 1.24).

ACE2 has diversified activities like regulation of angiotensin II (ANGII) which is responsible for increasing blood pressure and inflammation, causing damage to blood vessels linings and different types of tissue injury. ACE2 reduces the harmful effect of ANGII present in the epithelial cells of host cells. When SARS-CoV-2+ACE2 form a complex and prevent ACE2 from performing its normal function to regulate ANGII signaling. As a result, more ANGII are available and responsible for causing injury, especially to the lungs and heart, in COVID-19 patients. ANGII can increase inflammation and the death of cells in the alveoli which are critical for bringing oxygen into the body; these harmful effects of ANGII are reduced by ACE2 (Fig. 1.25). The sequential steps involved in pathway of coronavirus (COVID-19) are as follows (Fig. 1.26).

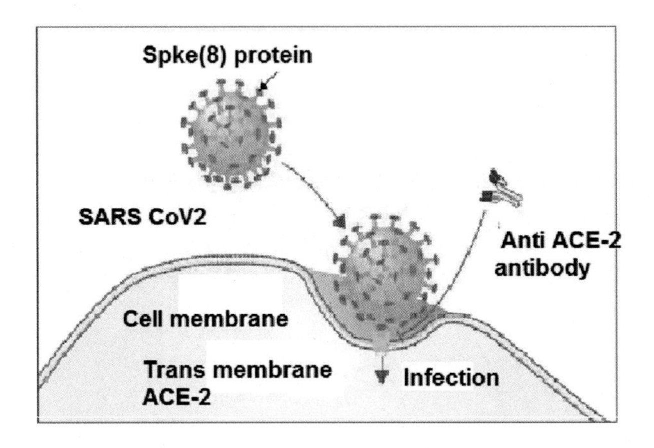

FIG. 1.24

Showing how ACE2 acts as the receptor for the SARS-CoV-2 virus and allows it to infect the cell.

Step 1

The virion attaches to the ACE2 present on the surface of host cells. The nucleocapsid plays a critical role in enhancing the efficiency of virus transcription and assembly. The nucleocapsids protect the genome and ensure its timely replication and reliable transmission.

Step 2

The virion + ACE2 enters the cytoplasm by fusion of the membrane of host cells and virion surface protein (S).

Step 3

The virion gets into the lysosome, and by the interaction with liposomal enzymes it releases genomic ssRNA and nonorganized capsid proteins.

Step 4

The genome RNA (+ sence) and capsid get into the cytoplasm.

Step 5

The nonstructural proteins get released. The N protein is thought to bind the genomic RNA in a beads-on-a-string fashion on the surface of the endoplasmic reticulum. It plays a key role in improving the efficacy of virus transcription and assembly.

Step 6

The genomic ssRNA and beaded proteins migrate toward the Golgi body.

Step 7

Release of progeny virion.

Here, the translation of the viral positive-sense single-stranded RNA (+ ssRNA) and cleavage of the translation product into specific viral protein occur. The entire process is carried by liposomes.

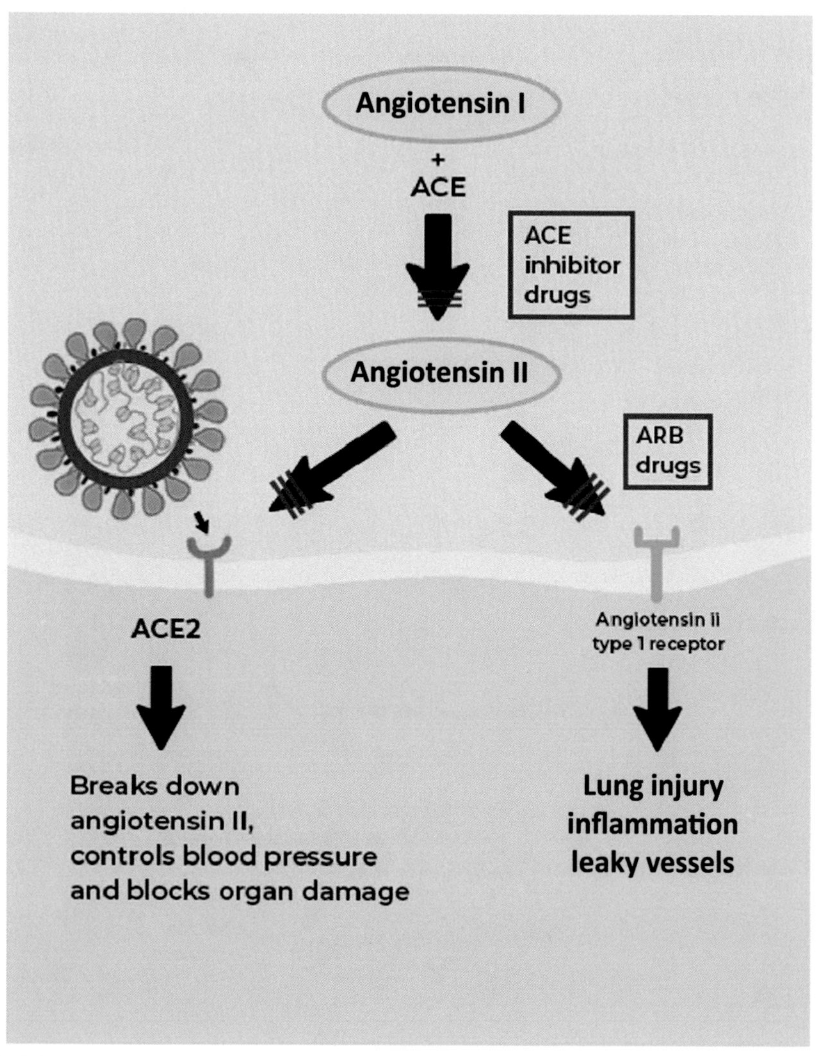

FIG. 1.25

ANGII increase inflammation and death of cells in the alveoli.

As discussed above, coronaviruses (COVID-19) have a high potential for genomic nucleotide substitution rates and recombination. The COVID-19 virus continues its propagation through host cells by keeping its genome expression system in intact form along with virus particles assembly and virion progeny release. Research is in progress to develop the safest and secure vaccine as a preventive measure for COVID-19 contamination.

In COVID-19, the rate of error during RNA replication (about 10^{-4}) is greater than that of DNA (about 10^{-5}). The RNA virus neither has the proofreading

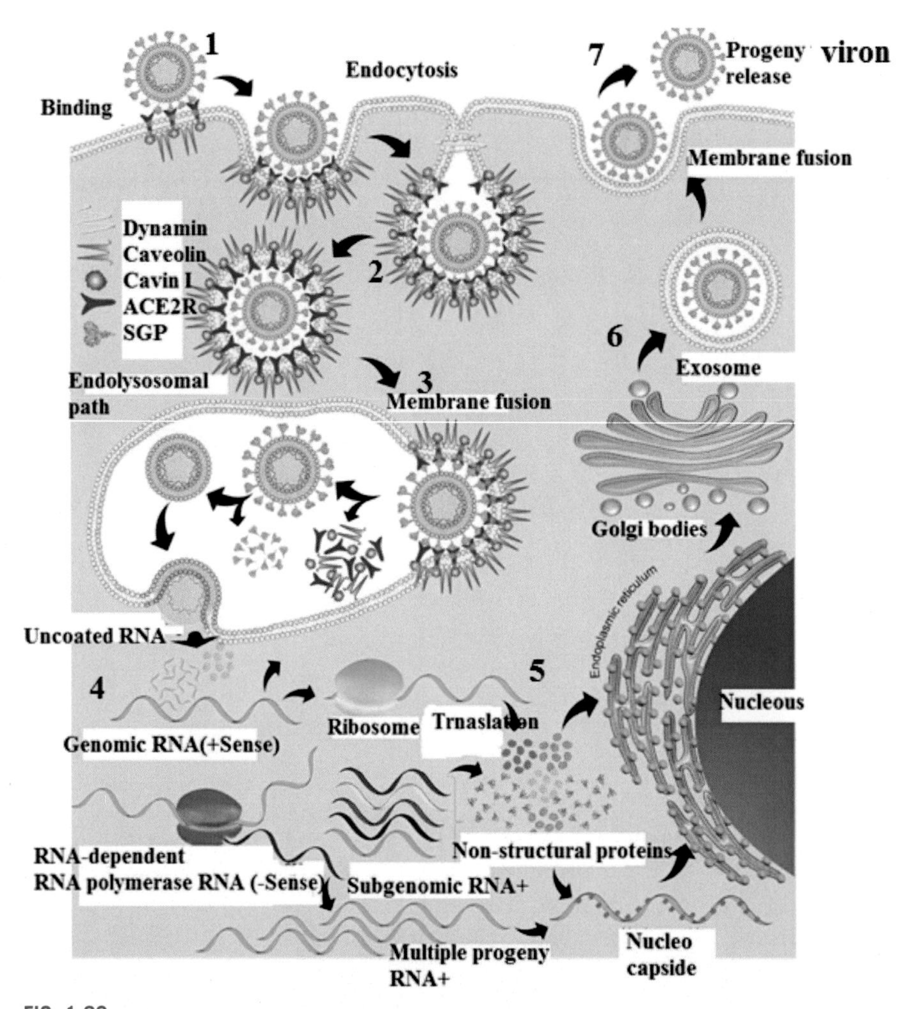

FIG. 1.26

The various steps involved in path way of coronavirus (COVID-19).

capabilities nor postreplication mismatch repair mechanism as one can do with DNA polymerase. So, the potential for mutation per replication cycle of an RNA genome is high [41]. Due to high mutation and recombination rates, coronavirus (COVID-19) can easily cross species barriers and adapt to new hosts [42].

References

[1] Detels R, Beaglehole R, Lansang MA, Gulliford M. Oxford textbook of public health. Oxford University Press; 2011.

[2] Jones J, Hunter D. Consensus methods for medical and health services research. BMJ 1995;311(7001):376–80.
[3] Jünger S, Payne SA, Brine J, Radbruch L, Brearley SG. Guidance on conducting and reporting DElphi studies (CREDES) in palliative care: recommendations based on a methodological systematic review. Palliat Med 2017;31(8):684–706.
[4] Aboud FE, Singla DR. Challenges to changing health behaviours in developing countries: a critical overview. Soc Sci Med 2012;75(4):589–94.
[5] Almirall D, Nahum-Shani I, Sherwood NE, Murphy SA. Introduction to SMART designs for the development of adaptive interventions: with application to weight loss research. Transl Behav Med 2014;4(3):260–74. https://doi.org/10.1007/s13142-014-0265-0.
[6] Bacon SL, Lavoie KL, Ninot G, Czajkowski S, Freedland KE, et al. An international perspective on improving the quality and potential of behavioral clinical trials. Curr Cardiovasc Risk Rep 2015;9(1):427.
[7] Gorski D. "Germ theory denialism: a major strain in "alt-med" thought". Science-Based Medicine; 2010. 08-09.
[8] Madigan MT, Martinko JM. Brock biology of microorganisms. Pearson Prentice Hall; 2006. ISBN: 978-0132017848.
[9] Goudsblom J. Public health and the civilizing process. Milbank Q 1986;64(2):161–88.
[10] Hanlon G, Pickett J. Public health administration andpractice. Times Mirror/Mosby; 1984.
[11] Grob GN. The state and the mentally ill: a history of Worcester State Hospital in Massachusetts, 1830–1920. Chapel Hill, N.C: University of North Carolina Press; 1966.
[12] Starr P. The social transformation of American medicine. New York: Basic Books, Inc; 1982.
[13] Turner JB, editor. Encyclopedia of social work. 17th ed. Washington, D.C.: National Association of Social Workers; 1977. editor in chief.
[14] Shereen MA, Khan S, Kazmi A, Bashir N, Siddique R. COVID-19 infection: origin, transmission, and characteristics of human coronaviruses. J Adv Res 2020;24:91–8.
[15] World Health Organization. Modes of transmission of virus causing COVID-19: implications for IPC precaution recommendations: scientific brief, 27 March 2020 (No. WHO/2019-nCoV/Sci_Brief/Transmission_modes/2020.1). World Health Organization; 2020.
[16] Winslow CEA. In: The evolution and significance of the modern public health campaign. South Burlington, VT: Journal of Public Health Policy; 1923. p. 1877–957.
[17] Fee E. Disease and discovery: a history of the Johns Hopkins school of hygiene and public health 1916–1939. Baltimore: Johns Hopkins University Press; 1987.
[18] Bremner RH, editor. Children and youth in America: a documentaryhistory. Cambridge, Mass: Harvard University Press; 1971.
[19] News item WHO reveals leading causes of death and disability 2019 https://www.who.int.
[20] UNAIDS 2019'AIDSinfo' [accessed July 2019].
[21] UNAIDS. Communities at the centre: global AIDS update 2019; 2019. p. 188 [pdf].
[22] UNAIDS. AIDSinfo; 2019 [accessed July 2019].
[23] Disease clusters: an overview case definition from the United States Department of Health and Human Services http://www.atsdr.cdc.gov/HEC/CSEM/cluster/case_def.html.
[24] Thompson WW, Weintraub E, Dhankhar P, Cheng OY, Brammer L, Meltzer MI, et al. Estimates of US influenza-associated deaths made using four different methods. Influenza Other Respi Viruses 2009;3:37–49.
[25] Nair H, Brooks WA, et al. Global burden of respiratory infections due to seasonal influenza in young children: a systematic review and meta-analysis. Lancet 2011;378:1917–30.

[26] Zhou P, Yang X, Wang X, Hu B, Zhang L, et al. A pneumonia outbreak associated with a new coronavirus of probable bat origin. Nature 2020;579:270–3.
[27] Hou YJ, et al. SARS-CoV-2 reverse genetics reveals a variable infection gradient in the respiratory tract. Cell 2020;182:429–446.e14.
[28] Lauer SA, et al. The incubation period of coronavirus disease 2019 (COVID-19) from publicly reported confirmed cases: estimation and application. Ann Intern Med 2020. https://doi.org/10.7326/M20-0504.
[29] Chao CYH, Wan MP, Sze To GN. Transport and removal of expiratory droplets in hospital ward environment. Aerosol Sci Tech 2008;42:377–94.
[30] Huang C, Wang Y, Li X, Ren L, Zhao J, Hu Y, et al. Clinical features of patients infected with 2019 novel coronavirus in Wuhan, China. Lancet 2020;395:497–506.
[31] Zhu N, Zhang D, Wang W, Li X, Yang B, Song J, et al. A novel coronavirus from patients with pneumonia in China, 2019. N Engl J Med 2020;382:727–33.
[32] Donoghue M, Hsieh F, Baronas E, Godbout K, Gosselin M, Stagliano N, et al. A novel angiotensin-converting enzyme-related carboxypeptidase (ACE2) converts angiotensin I to angiotensin 1-9. Circ Res 2000;87:E1–9.
[33] Patel S, Rauf A, Khan H, Abu-Izneid T. Renin-angiotensin-aldosterone (RAAS): the ubiquitous system for homeostasis and pathologies. Biomed Pharmacother 2017;94:317–25.
[34] Santos R, Sampaio W, Alzamora A, Motta-Santos D, Alenina N, Bader M, et al. The ACE2/angiotensin-(1-7)/MAS axis of the renin-angiotensin system: focus on angiotensin-(1-7). Physiol Rev 2018;98:505–53.
[35] Keidar S, Kaplan M, Gamliel-Lazarovich A. ACE2 of the heart: from angiotensin I to angiotensin (1-7). Cardiovasc Res 2007;73:463–9.
[36] Kiyono H, Fukuyama S. NALT-versus Peyer's-patch-mediated mucosal immunity. Nat Rev Immunol 2004;4:699–710.
[37] Brandtzaeg P, Kiyono H, Pabst R, Russell MW. Terminology: nomenclature of mucosa-associated lymphoid tissue. Mucosal Immunol 2008;1:31–7.
[38] Tacchi L, et al. Nasal immunity is an ancient arm of the mucosal immune system of vertebrates. Nat Commun 2014;5:1–11.
[39] Takeda K, et al. Allergic conversion of protective mucosal immunity against nasal bacteria in patients with chronic rhinosinusitis with nasal polyposis. J Allergy Clin Immunol 2019;143:1163–75.
[40] Neuman BW, Buchmeier MJ. Supramolecular architecture of the coronavirus particle. Adv Virus Res 2016;96:1–27. https://www.ncbi.nlm.nih.gov/pmc/articles/PMC7112365/.
[41] Rosenberg R. Detecting the emergence of novel, zoonotic viruses pathogenic to humans. Cell Mol Life Sci 2015;72:1115–25. https://link.springer.com/article/10.1007/s00018-014-1785-y.
[42] Lau SKP, Chan JFW. Coronaviruses: emerging and re-emerging pathogens in humans and animals. Virol J 2015;12:209.

CHAPTER

Prespective of universal health coverage

2

2.1 Introduction

Universal health care is also named as "universal health coverage" or "universal care." The basic target of universal health coverage (UHC) is to make sure that all people of a particular country or region get the opportunity for primary health services without any financial burden by paying for easy health insurance (Fig. 2.1).

In order to successfully launch this program a government should have adequate budgetary provision and well-trained health professionals (doctors, nurses, technical assistants, and health councilors). WHO has defined universal health coverage as: "to provide all people with access to needed health services (including prevention, promotion, treatment, and rehabilitation) of sufficient quantity to be effective, and ensure that the use of these services does not expose the user the financial hardship" (Fig. 2.2).

The overall strategies of UHC are to provide everyone access to the services that address the most significant causes of diseases and death and ensure that the quality of those services is good enough to improve the health of the people who receive them.

In order to have an action plan for UHC, in 2012, Sustainable Development Goals (SDGs) were proposed at the United Nations Conference organized in Rio de Janeiro (Fig. 2.3).

The main target was to develop an integrated approach that can meet the political, economic, and environmental challenges in order to develop a highly effective and efficient program related to universal health coverage. The entire section SDG 3 is dedicated to health-care coverage at the global level (Fig. 2.4).

The SDG 3 has 13 targets and 28 indicators to measure progress toward targets (Fig. 2.5).

The first nine targets are: (1) reduction in maternal mortality, (2) ending all preventable deaths under 5 years of age, (3) fight communicable diseases, (4) ensure reduction in mortality from noncommunicable diseases, (5) promote mental health, (6) prevent and treat substance abuse, (7) reduce road injuries and death, (8) grant universal access to sexual and reproductive care, and (9) family planning and education, and achieve universal health coverage and reduce illness and death from hazardous chemicals and pollution.

The four "means for achieving" SDG 3 targets are (10) implement the WHO Framework Convention on Tobacco Control, (11) support research, (12) development

Healthcare Strategies and Planning for Social Inclusion and Development. https://doi.org/10.1016/B978-0-323-90446-9.00002-2

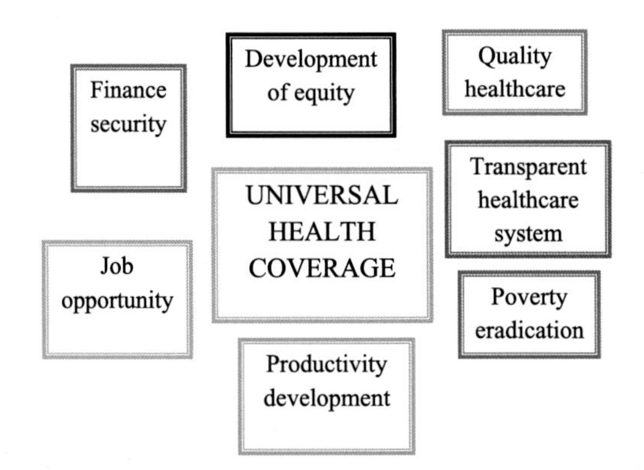

FIG. 2.1

Scope of universal health coverage.

FIG. 2.2

Conceptual redevelopment of model for universal health.

Source: Universal health coverage who.int.

FIG. 2.3

Targets sustainable development goals (SDGs) was proposed at the United Nations Conference organized at Rio de, 2012.

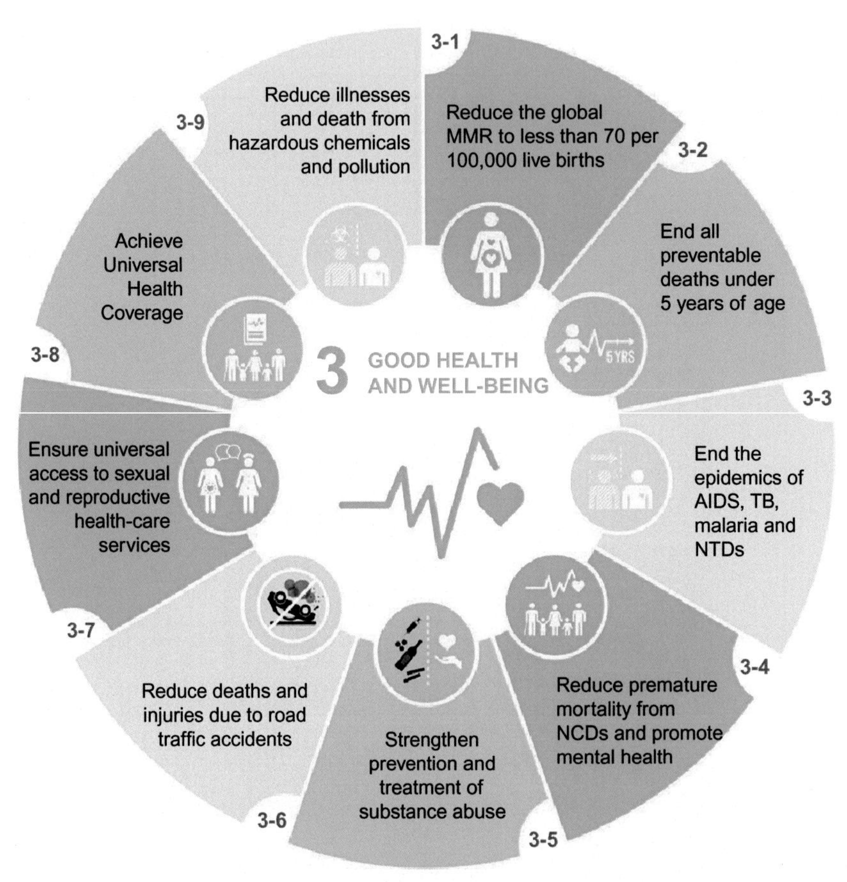

FIG. 2.4

Sustainable development goal 3 related to health coverage at the global level.

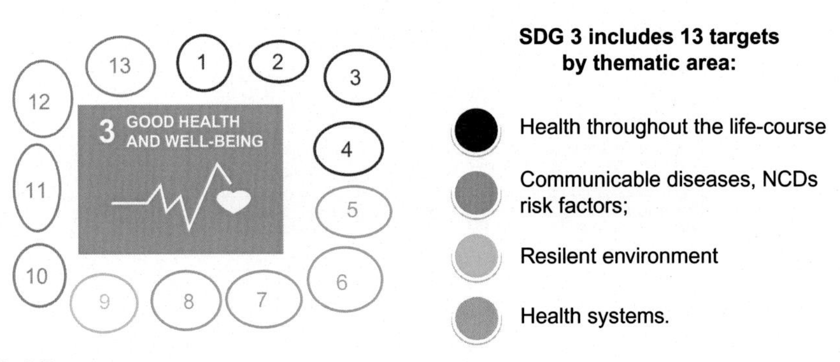

FIG. 2.5

Detail on SDG 3 goal with 9 target points and 4 points on means to achieve. All the 13 points are described in detail in the text.

and universal access to affordable vaccines and medicine, and (13) increase health financing and support health workforce in developing countries.

Meanwhile, remarkable progress has been made to increase life expectancy and reduction in the common causes of child and maternal mortality. Between 2000 and 2016, the worldwide under-5 mortality rate decreased by 47% (from 78 deaths per 1000 live births to 41 deaths per 1000 live birth). Still, the number of children dying under age 5 is very high: 5.6 million in 2016 [1].

In 2015, for a better workout, the SDGs replaced the Millennium Development Goals (MDGs) which started a global effort in 2000 to eradicate poverty and hunger and to have universal health coverage (Fig. 2.6) [2–5].

2.2 Move toward universal health coverage

Since the last two decades, many countries have been showing interest in adopting universal health coverage as national policy, which is mainly based on: (i) the population that is covered by pooled funds, (ii) the proportion of direct health cost covered by pooled funds, and (iii) the health services covered by those funds.

Looking at the growing interest for a systematic assessment, Japan and the World Bank started a 2-year collaborative global survey to understand global status on health coverage [6]. In this program, 11 countries (Bangladesh, Brazil, Ethiopia, France, Ghana, Indonesia, Japan, Peru, Thailand, Turkey, and Vietnam) with diversified lifestyles were involved. The main target was to find a proper way out to successfully move toward UHC.

FIG. 2.6

Millennium Development Goals with 8 target points. Target 5 highlights health coverage.

2.3 UHC goals and implementation

To eradicate poverty and ensure prosperity, all countries need a sustainable, inclusive development strategy based on human capital investments in health, education, and social protection for all. The last few decades have witnessed, for example, how during financial crisis in Indonesia, Thailand, and Turkey; at the time of redemocratisation in Brazil; and during post-World War II reconstruction effort in France and Japan, UHC served as vital mechanism for improving the health and welfare of their citizen, while developing the foundation for economic growth and competitiveness grounded on the principles of equality and sustainability. Even, before 1999s, a lot of countries in Latin America, the Caribbean, Africa, and the Asia-specific region, including China and Brazil that has the largest universal health care system in the world, took steps to bring their populations under universal health coverage [7, 8].

In 2018, India introduced a tax-payer funded decentralized universal health-care system that could reduce mortality rates and malnutrition.

Survey report, in 2012, focused that Ghana, Rawanda, Nigeria, Mali, Kenya, Indonesia, the Philippines, and Vietnam that show a lot of progress have been made in the implementation of UHC [9, 10]. At present, most industrialized countries and many developing countries operate some form of publicly funded health care with universal health coverage as the primary goal. However, according to the National Academy of Medicine and others, United States is only a wealthy, industrialized nation that does not provide universal health care [11, 12].

2.4 UHC and SDG 3

The SDG 3.8.1 and SDG 3.8.2 are mainly targeted to proportion of a population that can access essential quality health services, and the proportion of the population that spends a large amount of household income on health, respectively.

At present, globally, people are in trial toward UHC, even though COVID-19 is a big threat as pandemic responsible for extremely unavailability of health services. So, it is high time that all countries should, jointly, take well-planned actions to eradicate COVID-19, or to maintain the gains they have already made. Moving toward UHC for a country needs stable health systems and robust financing structures to monitor key factors. Financially poor people face problems in normal life, as they spend their pocket money for health coverage, and even the rich may be exposed to financial hardship in the event of severe or long-term illness, across a population.

The improvement of health service coverage and health outcomes depends on the availability of well-trained health workforce, accessibility, and delivery of quality services timely. The COVID-19 pandemic demonstrated an excellent example of the invaluable role of the health care workforce and the importance of investments in this area. As suggested by SDGs and UHC targets about 18 million additional health workers are needed by 2030. But, overall demands for health workers are concentrated in low- and lower-middle-income countries. Looking at the present situation, the growing demand for health workers is projected to add an estimated 40 million health sector jobs to the global economy by 2030.

2.5 Tracking UHC

In 2011, The World Bank, jointly with WHO, has developed a framework to track the progress of UHC, in a meeting held at Geneva. The main target of the meeting was to prepare a draft to work out and track universal coverage, delivery of health care, health-care financing, health services accessibility, and program evaluation (Source: www.who.int).

Fortunately, the UHC movement has gained global momentum, with the first-ever UN high-level meeting on UHC held in September 2019. A Political Declaration was unanimously adapted member States, affirming their high-level political commitment to UHC and outlining a number of necessary actions. Twelve cosignatories including the WBG (World Bank Group's) also launched the Global Action Plan for Healthy Lives and Well-being for All (GAP) to jointly support countries delivering on the SDG 3 target. In January 2020, the second UHC Forum was held in Bangkok, targeting to enhance political momentum on UHC, at the global level.

Achieving universal health coverage, including essential services coverage and financial protection for all, is target 3.8 of the Sustainable Development Goals (SDGs).

As a part of SDG goals: (i) UHC also focuses on health services as essential for the development of a country. The UHC system provides a platform for an integrated approach within the health sector. (ii) The SDGs and UHC are intrinsically about improving equity. Polices, programs, and monitoring should focus on progress among the poorest people, women and children, and people living in rural areas and from monitoring groups. Using, UHC as a common monitoring platform ensures a continuous focus on health equity. (iii) The health goal is closely linked to many of the other social, economic, and environmental SDGs. Intersectoral action, including a major emphasis on pro and prevention, is urgently needed. To end poverty and boost shared prosperity, countries need to robust inclusive economic growth and to drive growth, and they also need to build human capital through investments in health, education, and social protection for all their citizens. In order to free the world from extreme poverty by 2030, countries must ensure that all their citizens have access to quality, affordable health services.

For framing UHC tracking strategies, a country needs strong, transparent monitoring and review system, as well as regular implementation and service delivery search that jointly feeds an ongoing learning process of UHC implementation. It is also necessary that both health information systems and the science of service delivery require more investment if results are translated into targeted action.

It is essential to have some essential guidelines related to indicators to assess and understand the progress of UHC, after its implementation. The World Health Report (2010) outlined a conceptual framework for UHC (World Health Organization (WHO, 2010) that suggested three broad dimensions: (i) the range of services that covered (service coverage), (ii) proportion of the total costs covered through insurance or other risk pooling mechanism, and (iii) the proportion of the population covered (population coverage).

The processes for universal health coverage involve important policy choices and inevitable trade-offs. The cumulated funds obtained through government budgets, compulsory insurance contribution (payroll taxes), and household and/or employer prepayments for voluntary health insurance are organized and used for UHC (Fig. 2.7).

Pooled funds can also be used for coverage to those individuals who previously were not covered to services that previously were not covered or to reduce the policy direct payments needed for each service. These dimensions represent a set of policy choices about benefits and their rationing that are among the critical decisions facing countries in their reform of health financing systems toward universal coverage.

For the sake of convenience and understanding the efficacy of UHC, WHO.

WHO recommended 16 important indicators under four categories:

(i) Reproductive, maternal, newborn, and child health:
- family planning
- antenatal and delivery care
- full child immunization
- health-seeking behavior for pneumonia

(ii) Infectious diseases: tuberculosis treatment
- HIV antiretroviral treatment
- use of insecticide-treated bed nets for malaria prevention
- adequate sanitization

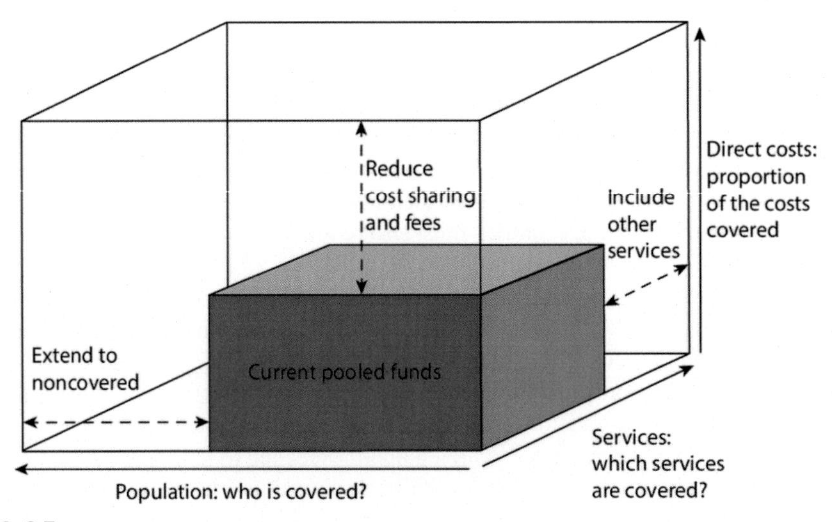

FIG. 2.7

Three dimensions of the universal health coverage.

(iii) Noncommunicable diseases
- prevention and treatment of raised blood pressure
- cervical cancer screening
- tobacco (non) smoking

(iv) Service capacity and access
- basis hospital access
- health workers density
- access to essential medicines

The framing and implementation of UHC depends on their requirement and financial status. But there is also value in a global approach that uses standardized measures that are internationally recognized so that they are comparable across borders and over time.

The UHC is mainly based on 1948 WHO human right and commits to ensuring the highest attainable level of health for all. WHO supports countries to develop their health systems to move toward and sustain UHC and to monitor progress. For this purpose, WHO works with many different partners in different situations and for different purposes to advance UHC around the world. Some of the WHO's partnerships include:

- UHC 2030
- Alliance for health policy and system research
- P4H social health protection network
- UHC partnership
- Primary health-care performance initiative

From October 25 to 26, 2018, WHO in partnership with UNICEF and the Minister of Kazakhstan hosted the Global Conference on Primary Health Care, 40 years after the adaptation of the historic Declaration of Alma-Ata. Ministers, health workers, academics, partners, and civil society came together to recommit to primary health care as the keystone of UHC in the bold new Declaration of Astana. The declaration aims to renew political comment to primary health care from government, nongovernment organization, professional organizations, academia, and global health and development organizations.

2.6 Funding model

Universal health coverage does not represent the coverage of all cases and for all people. Funding system for health care may be from the government or from health-care insurance based on a requirement that all citizens purchase private health insurance. World Health Organization emphasizes without involving the public without financial hardship.

2.6.1 History

In 1883, Germany had taken credit by launching a national health insurance system, with the Sickness Insurance Law which was collected from the deduction in workers' wages and from employer's contribution. At a later stage, other countries started following the same type of insurance funding generating system. In the United Kingdom, the National Insurance Act 1911 provided coverage for primary care (by not a specialist or hospital care) for wage earners, covering about one-third of the population. In 1912, the Russian Empire established a similar system, and other industrialized countries began to follow the same pattern of raising funds. By the 1930s, all of Western and Central Europe and Japan adopted the same type of funding development system and expanding further upon in 1935 and 1940. Following the Russian Revolution of 1917, the Soviet Union established a fully public and centralized health care system in 1920.

From 1939 to 1941, New Zealand created a health care system in a series. In 1949, Australia, the state of Queensland, introduces a free public hospital system. Following World War II, universal health care systems began to be set up around the world. On July 5, 1948, the United Kingdom launched its universal National Health Service. At a later stage, universal health care was introduced in the Nordic Countries of Sweden (1955), Iceland (1956), Norway (1956), Denmark (1961), and Finland (1964) [13–16].

In 1961, Japan introduced health care, and in Canada, it was implemented state wise. In 1050 and 1957, Kuwait and Bahrain, respectively introduced a healthcare system. Italy introduced its Servizio SAnitorio Nationalle (National Health Service) in 1978. In 1975, Australia implemented universal health insurance with the Medibank, which led to universal coverage under the current Medicare system from 1984.

South and Western Europe, from 1970 to 2000, began to introducing universal coverage, based on earlier health insurance programs. For example, France built upon its 1928 national health insurance system and covered health insurance for about 1% of the population who had been suffering due to lack of health insurance policy since 2000.

Finland (1972), Portugal (1979), Cyprus (1980), Spain (1986), and Iceland (1990) introduced a universal health system based on an insurance mandate in 1994. Some Asian countries including South Korea (1989), Taiwan (1995), Singapore (1993) Israel (1995), and Thailand (2001) introduced universal health coverage from the general public.

Following the collapse of the Soviet Union, Russia retained and reformed its universal health care system.

Latin America, the Caribbean, Africa, and Asia-Pacific, currently have a universal health coverage system. China and Brazil are the world's largest universal health coverage systems which improved 80% of the population. India introduced a taxpayer funded decentralized healthcare system that helps reduce mortality rates and malnutrition.

Presently, industrialized countries prefer to adapt public-funded health care with the universal coverage goal. According to the National Academy of Medicine and others, the United States is the only wealthy, industrialized nation that does not provide universal health care.

2.6.2 Compulsory health insurance

A wide range of compulsory insurance is available for the public and business either as a mandatory or compulsory system. Compulsory insurance is supposed to protect accident victims against the costs of recovering from an accident someone lease, such as another driver or an employer has caused. Compulsory health insurance is often provided by the government having the choice of multiple public and private funds providing a standard service (as in German) or sometimes just a single public fund (as in the Canadian provinces). Health care in Switzerland is based on compulsory insurance.

Some European countries (Germany, Belgium, Netherland) provide insurance and universal health care. In this type of coexistence system, the problem of the adverse pool is made to equalize risk between funds.

In such a funding system, the younger population play a compensation pool, and the older population receives funds from the pool. In this way, the population on high risk are protected compensation on sickness funds by risk-adjusted capital payments. The health funding system of a country, mainly, dependent on the condition of a country on which basis the health funding system is operated.

For example, the Republic of Ireland at one time had a "community" system by Vhi Healthcare (VHI), effectively a single payer or common risk pool. Still, the Vhi is the largest health insurance company in Ireland. It is a statutory corporation whose members are appointed by the ministry of health. It is regulated by the Health Insurance Authority.

In Poland, people pay a percentage of the average monthly wage to the state, even if they are covered by private insurance. People, under the employment contract, pay a percentage of their wage, while entrepreneurs pay a fixed rate, based on the average national wage. Unemployed people are insured by the labor office [17–19].

2.6.3 Single payer

As stated above, "universal coverage" refers to a health care system where every individual has health coverage. On the other hand, a "single-payer system" is one in which there is one entry usually by the government. "single-payer" system is a funding mechanism and refers to health care financed by a single public body from a single fund and does not specify the type of delivery or for whom doctors work. Although the fund holder is usually the state, some forms of single payer use a mixed public-private system.

2.6.4 Tax-based financing

In 2004, a meeting held on to introduce tax-based health system has emerged as a new concept on raising health funding system through tax money [20]. Until the mid-20th century, the major insurance funding system from out-of-pocket payment for healthcare services were private philanthropies, mutual association, or social insurance plans (sickness fund). The government used to be responsible for meeting hospital health care until the 20th century.

Germany has taken first credit to combine its sickness funds into a social health insurance coverage, dated from the second half of the 19th century.

Tax-based payment system, through the government, has two different ways: (i) the tax-based system by earlier development of social or private health insurance. For example, Britain passed its National Insurance Act in 1911, financed through payroll contributions and did not adopt a universal tax-supported health system until after World War II. This pattern is common among Western European countries, and (ii) the tax-based system evolved from health services administered directly by colonial regimes. This pattern is found mainly among developing countries that were colonized or heavily influenced by Britain such as Malaysia, Singapore, Hong Kong, and many countries in Africa and the Caribbean.

The tax-based payment system

The tax-based systems for raising funds for universal health care have both advantages and disadvantages. Since payment is mandatory, the system avoids many problems that are common to voluntary insurance markets. Tax-based system can benefit from scale economies in administration, risk management, and purchasing power. These strengths come from the collective and political nature of raising and allocating tax revenues in a modern nation-state. In general, tax-based financing serves contribute to providing health services through various taxes. Some countries including United Kingdom, Ireland, New Zealand, Italy, Spain, Brazil, Portugal India, and the Nordic countries have the option to raise healthcare funding system from taxation alone.

2.6.5 Social health insurance

Social health insurance (SHI) is a concept where the government intervenes in the insurance market form of financing managing health care based on risk pooling. This is done through a process where individual's claims are partly dependent on their contributions, which can be considered as insurance premiums to create a common fund out of which the individuals are then paid benefit in the future. The SHI pool system is mainly based on the health risk of the people on one hand, and the contribution of individuals, household, enterprise, and the government on the other hand.

Japan and the Republic of Korea are among the first countries in Asia and Pacific, to adapt SHI. While lower middle-income countries such as Thailand and the Philippines have a high proportion of SHI coverage. Developing countries with stronger economies such as China, Indonesia, and India have lower population coverage through SHI Schemes. Implementation of SHI, mainly, depends on the level of socioeconomic development, financial sector development (mainly banking), and employment conditions, especially the existence of a larger proportion of formal sector-organized establishment. The successful implementation of SHI is mainly dependent on national consensus on the policy framework, strong regulation, and adequate administrative capacity.

The 48th session of the WHO Regional Committee for South-East Asia held in Colombo in September 1995 discussed the issue of alternative healthcare financing reforms, within the framework of solidarity, equity, and expanding essential coverage. At the later stage, after many subsequent meet, it was decided that financial reformations through an update of national health accounts and related surveys are needed for successful implementation of SHI.

Examples of social insurance include:

Medicare

Medicare is a national health insurance program in the United States, started in 1965 under the Social Security Administration (SSA) and now administered by the Centres for Medicare Services (CMS). It primarily provides health insurance for Americans aged 65 and older, but also some younger people with disability status as determined by the SSA, and people with end-stage renal disease and amyotrophic lateral sclerosis (ALS or Lou Gehrig's disease).

Social security

Social-security is a commonly used terminology in the United States. The original social security act was signed into law by Franklin D. Rooevelt in 1935 and encompasses several social welfares and social insurance programs.

The Federal Insurance Contributions Act Tax (FICA) or Self Employed Contributions Act Tax (SECA) is involved in raising social security fund and are formally entrusted to the Federal Old-Age and Survivors Insurance Trust Fund and Federal Disability Insurance Trust Fund.

Unemployment insurance

Unemployed payment benefits, also known as unemployment insurance, unemployed payment, unemployment compensation, or simply unemployment. In the United States, people get benefits depending on the status of a person and jurisdiction. Unemployment benefits are generally given only to those registering as unemployed, and often on conditions ensuring that they seek work.

2.6.6 Private insurance

Private health insurance refers to any health insurance coverage that is offered by a private entity instead of a state or federal government. In private insurance system, the premium is directly deducted from employers, associations, individuals, and families to insurance companies, which pool risks across their membership base. Generally, private insurance policies are sold by commercial for-profit firms, nonprofit companies, and community health insurers.

In some countries with universal coverage, private insurance often bears expensive and state healthcare system support for insurance payment. For example, BUPA, the United Kingdom healthcare system has the highest coverage policy, most of which are routinely provided by the National Health Service. In the United States, dialysis treatment for end-stage renal failure is generally paid for by the government and not by the insurance company.

The private health insurance system is either direct or indirect way of the collection of money to raise funds for health services. In a direct way, an insurance payer spends money from out-of-pocket payments. In private fund-raising system, the government has limited resources to raise enough resources to finance health services. The indirect (overhead) costs would include, but not be limited to, general support staff and related costs, insurance, taxes, floor space, facilities, and administration.

In the private health system, the competition and ownership incentives offer the public an opportunity for the option of a better system for efficient health services. However, instability of the private system sometimes brings dilemma in public to have the option of private health services. Except for a few countries such as Switzerland and the United States, private health insurance is a minor source of finance. This is mainly due to the people from developing countries cannot afford private insurance. It has been observed that most of the industrialized countries like to have the option for the free market to produce and allocate health care in general and mental health care in particular. Government intervention improves the unnecessary monopoly of the private healthcare system. It is necessary to have monitor on public regulation of private markets to ensure effective risk pooling, affordability, informed choice, and continuity of coverage [21–24].

2.6.7 Community-based health insurance

Community-based health insurance (CBHI) systems are voluntary and designed by community members pooling funds to offset the cost of health care. But, the impact of CBHI on financial protection and access to needed health care are moderate. So, it has been observed that poor people usually remain excluded from such a system. However, the community healthcare system is accountable for health-care providers.

The main drawback with CBHI is its unreliability without any subsidization of poor and vulnerable groups. CBHI system can only play a complementary role rather than provide any benefit toward UHC. Keeping in view on universal health coverage, WHO supports the Member States to develop health financing strategies that aim at reducing fragmentation and better pooling to enhance the potential for redistributive capacity [25–27].

2.7 Global strategies for UHC

The universal health coverage refers to access to health services without financial hardship. It includes the full range of essential health services, from health promotion to prevention, treatment, rehabilitation, and palliative care.

At present, about half of the world population do not have adequate health services. About 100 million people are pushed into extreme poverty each year because of out-of-pocket spending on health. In order to have a successful implementation, it is necessary to realize the government and its efficacy in the working capacity of skilled health workers for UHC services. Presently, Covid-19 is a burning example of how to challenge the pandemic global condition, and to bring stable, equitable, societies, and economies to the problem for an immediate solution. This crisis is an opportunity to seize the moment to make changes that benefit both UHC and health security.

References

[1] https://data.unicef.org, (2018) Progress for Every Child in the SDG Era—UNICEF DATA.

[2] World Health Organization. The world health report: health systems financing: the path to universal coverage. Geneva: World Health Organization; 2010, ISBN:978-92-4-156402-1.

[3] Matheson D. Will universal health coverage (UHC) lead to the freedom to lead flourishing and healthy lives? Comment on "inequities in the freedom to lead a flourishing and healthy life: issues for healthy public policy". Int J Health Policy Manag 2015;4(1):49–51.

[4] Abiiro GA, De Allegri M. Universal health coverage from multiple perspectives: a synthesis of conceptual literature and global debates. BMC Int Health Hum Rights 2015;1472-698X. 15:17. https://doi.org/10.1186/s12914-015-0056-9. PMC 4491257 26141806.

[5] Universal health coverage (UHC). World Health Organization; 2016. December 12, 2016.

[6] Maeda A, Araujo E, Cashin C, Harris J, Ikegami N, Reich M. Universal health coverage for inclusive and sustainable development: A synthesis of 11 country case studies. Washington, DC: World Bank; 2014.

[7] Yu H. Universal health insurance coverage for 1.3 billion people: What accounts for China's success? Health Policy 2015;119(9):1145–52. and Brazil (24).

[8] Gómez EJ. In Brazil, health care is a right. CNN; July 13, 2012.

[9] Eagle W. Developing countries strive to provide universal health care. World Health Report; 2013.

[10] The World Bank. Universal Healthcare on the rise in Latin America; 2013.

[11] Insuring America's Health. Principles and Recommendations Archived 2007-08-18 at the Wayback Machine. Institute of Medicine at the National Academies of Science; 2004.

[12] The Case for Universal Health Care in the United States, cthealth.server101.com; 2018.

[13] Rowland D, Telyukov AV. Soviet healthcare from two perspectives. Health Aff 1991;10(3):71–86.

[14] OECD Reviews of Health Systems. OECD reviews of health systems: Russian Federation; 2012. p. 38.

[15] Abel-Smith B. Social welfare; social security; benefits in kind; national health schemes. The new encyclopedia Britannica. 15th ed. Chicago: Encyclopedia Britannica; 1987, ISBN:978-0-85229-443-7.

[16] Richards R. Two social security acts. Closing the door to destitution: the shaping of the Social Security Acts of the United States and New Zealand. University Park: Pennsylvania State University Press; 1993. p. 14, ISBN:978-0-271-02665-7.

[17] Varkevisser M, van der Geest S. Competition among social health insurers: a case study for the Netherlands, Belgium and Germany. Res Healthc Financ Manag 2002;7(1):65–84.

[18] Rothschild M, Stiglitz J. Equilibrium in competitive insurance markets: an essay on the economics of imperfect information. Q J Econ 1976;90(4):629–49.

[19] Belli P. How adverse election affects the health insurance market. Policy research working paper 2574, Washington, DC: World Bank; 2001.
[20] Savedeff W. Tax-based financing for health systems: options and experience. Department "Health System Financing, Expenditure and Resource Allocation" (FER) Cluster "Evidence and Information for Policy" (EIP); 2004.
[21] World Health Organization. Health financing mechanisms: private health insurance. Geneva: World Health Organization; 2008.
[22] Bupa. Individuals: health and life cover: health care select 1: key features of this health insurance plan: what's covered? What's not covered?; 2010.
[23] Centers for Medicare & Medicaid Services. Medicare coverage of kidney dialysis & kidney transplant services. Baltimore: Centers for Medicare & Medicaid Services; 2010.
[24] Varshney V, Gupta A, Pallavi A. Universal health scare. Down to earth. New Delhi: Society for Environmental Communications; 2012.
[25] Gray M, Pitini E, Kelley T, Bacon N. Managing population healthcare. J R Soc Med 2017;110(11):434–9.
[26] National Audit Office. International health comparisons: a compendium of published information on healthcare systems, the provision of health care and health achievement in 10 countries. London: National Audit Office; 2003.
[27] Grosse-Tebbe S, Figueras J. Snapshots of health systems: the state of affairs in 16 countries in summer 2004. Copenhagen: World Health Organization on behalf of the European Observatory on Health Systems and Policies; 2004.

CHAPTER

Public health, food security, and hunger 3

3.1 Introduction

Global public health system is a conglomeration of all organized activities for the prevention of disease, a sustainable healthy service, prolonging life, and for promoting health for all (Fig. 3.1) [1–6].

The World Health Organization defines public health as: "*the art and science of preventing diseases, prolonging life and promoting health through the organized efforts of society*" [7].

In general, health care is referred to the restoration of human life with safe and secure of body and mind. So, the medical dictionary defines health care as: "The *prevention, treatment, and management of illness and the preservation of mental and physical well-being through service offered by the medial and allied health profession.*"

So, in order to bring sustainable health-care practices, globally, the 2030 Agenda for 17 Sustainable Development Goals targeted to have poverty-free sustainable health-care practices all over the world by the end of 2030. This is in continuous of the Millennium Development Goals (MDGs) signed in September 2000. In this agenda, there are eight goals with 21 targets and a series of measurable health indicators and economic indicators for each target [8–10].

The main target of SDGs is to promote efficient health-care services and well-being in a sustainable pattern by integrating public health services and reducing social exclusion. The success in SDGs could only be possible by bringing the other sectors (mainly NGOs and health-care professionals) closer to address the wider determinants of health, as decided by the Alma-Ata Declaration. In 1978, the Alma-Ata Declaration has emerged as a major milestone of the 20th century in the field of public health to achieve the goal of health for all.

But, currently most of the developing countries are facing challenges that include:

- economic crisis
- lower education rate
- high level of birth rate
- high infant mortality rate
- poor infrastructure
- weak government policy
- poor access to health care

Healthcare Strategies and Planning for Social Inclusion and Development. https://doi.org/10.1016/B978-0-323-90446-9.00003-4

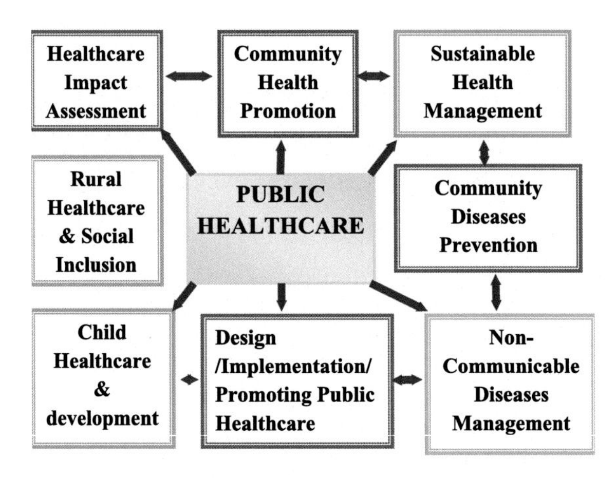

FIG. 3.1

Schematic representations of different aspects of health-care practices.

The wide gap existing in health-care services between developing and developed countries is mainly due to economic and political instability. The WHO has identified indicators that are used to measure the discrepancies in health such as: "health *of the population, fair financial contributions, and responsiveness of the system, preventable deaths, and affordability*?"

In 2018, the World Health Organization (WHO), United Nations International Children's Emergency Fund (UNICEF), and the Government of Kazakhstan cohosted the Global Conference on Primary Health Care in Astana. It was a mega gathering of world leaders, government ministers, development partners, civil society, and health-care professionals. The main agenda of the conference was to renew their commitment to primary health care as the means of achieving universal health coverage, SDG3, and other SDG goals related to health-care services at global level. The conference was held after 40 years since the first Global Conference on Primary Health Care, held in 1978 in Almaty (then Alma-Ata), Kazakhstan. It has been noticed that across the WHO European Region the main challenges prevailing around the country include unstable economic crisis, unstable government, environmental degradation, and natural disaster. These factors cumulatively or individually are responsible for irregularity and inadequate health services. Most of the developing countries have also been facing the same sort of hurdles while discharging health-care services. This is mainly due to changing in diseases patterns across the region, which ultimately leads to public health emergencies, changes in lifestyle patterns and increasing prevalence of noncommunicable diseases (NCDs), and emerging of communicable diseases. So, in order to mitigate the barriers related to health services, the WHO Regional Office for Strengthening Public Health Capacities and Services has adopted the European Action Plan (EAP) to secure the public health challenges.

Although health care is a burning issue for both the developed and developing countries, still, there are many hurdles acting as barriers while implementing health-care project granted by either through government or any international agency (Fig. 3.2). The issues related to health care are a chronic and crucial problem for a country irrespective of their financial status.

Design and development of policy is an initiation for taking care of health-care issues, but it is most necessary how effectively the policy matters are to be challenged for public welfare. Indirectly, the challenges to implementation are referred to as "implementation barriers (Fig. 3.2)."

Poverty is a major cause of ill health and a barrier to access health care when needed. The financially poor people cannot effort for better medicine for health care and proper dietary intake. That is why two goals of SDGs are targeted to develop "good health and well-being" (Fig. 3.3).

Following are different aspects of health-care practices in order to have successful "health for all" at the global level.

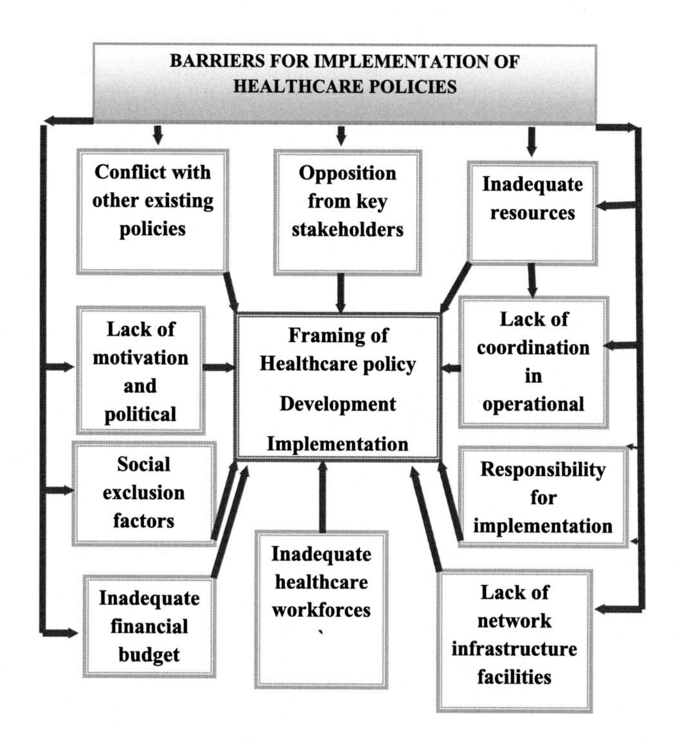

FIG. 3.2

Schematic presentation of different types of barriers for implementation and workout of healthcare policies.

FIG. 3.3

The first two SDGs goals are to implement the target for "health for all."

3.2 Community health promotion

In the process of overall health promotion and diseases prevention, it is advisable to entrust this responsibility to the individual level in order to reduce their risk of diseases and disability. This could be helpful in reducing or eliminating health disparities, improving health quality, and increasing the efficacy of health-care services. Community health promotion can be achieved in five different stages (Fig. 3.4).

But, before going ahead with implementing any program related to community health promotion, it is necessary to assess the different aspects of the community in detail. A community health assessment (CHA) is an important tool in understanding the structure and function of the community and necessity of promoting health services on a priority basis (Fig. 3.5).

Both quantitative and qualitative can be adapted for CHA to understand the health of the specific community. CHA is one of the best tools to revive data related to the

FIG. 3.4

The community health promotion in five different stages.

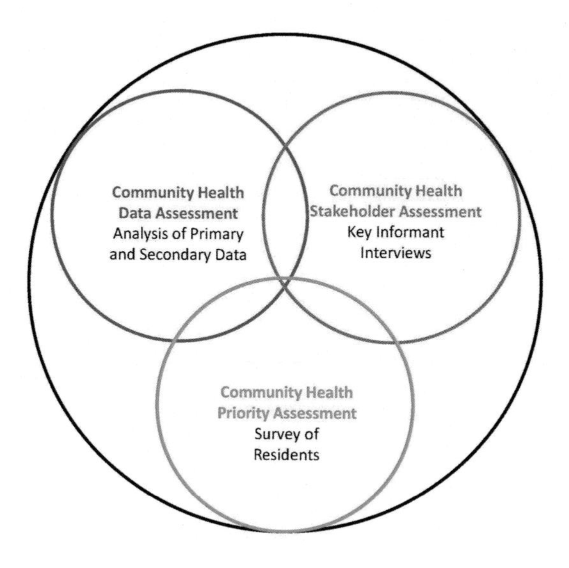

FIG. 3.5

Conceptual developments on community health assessment.

quality of life, morbidity (impact of living with illness), information on community assets, and social and economic factors that influence health and quality of life. This data bank can be well used broadly for community decision-making, prioritization of health problems, and the development, implementation, and evolution of community health improvement plans.

So, health promotion covers a wide range of activities including individual, community, and populations for the prevention of diseases through proper planning activities and programs for those who can take care of "health for all" projects. Health promotion can strengthen an individual to reduce the risk of diseases and disability. At the population level, it can play a critical role in exempting health disparities, improve quality of lifestyle, and improve the accessibility of efficient health services. Health promotion programs can be implemented, both in the developed and underdeveloped countries, including the rural and remote communities.

3.3 Sustainable health management

A societal health and welfare is a fundamental to have a sustainable pattern of healthy life, which has environmental, social, and economic impact on health-care system (Fig. 3.6).

The World Health Organization (WHO) defines an environmentally sustainable health system as: "*improve, maintains or restores health, while minimizing negative impact on the environmental and leveraging opportunities to restore and improve it, to the benefit of the health and well-being of current and future generation*" [11].

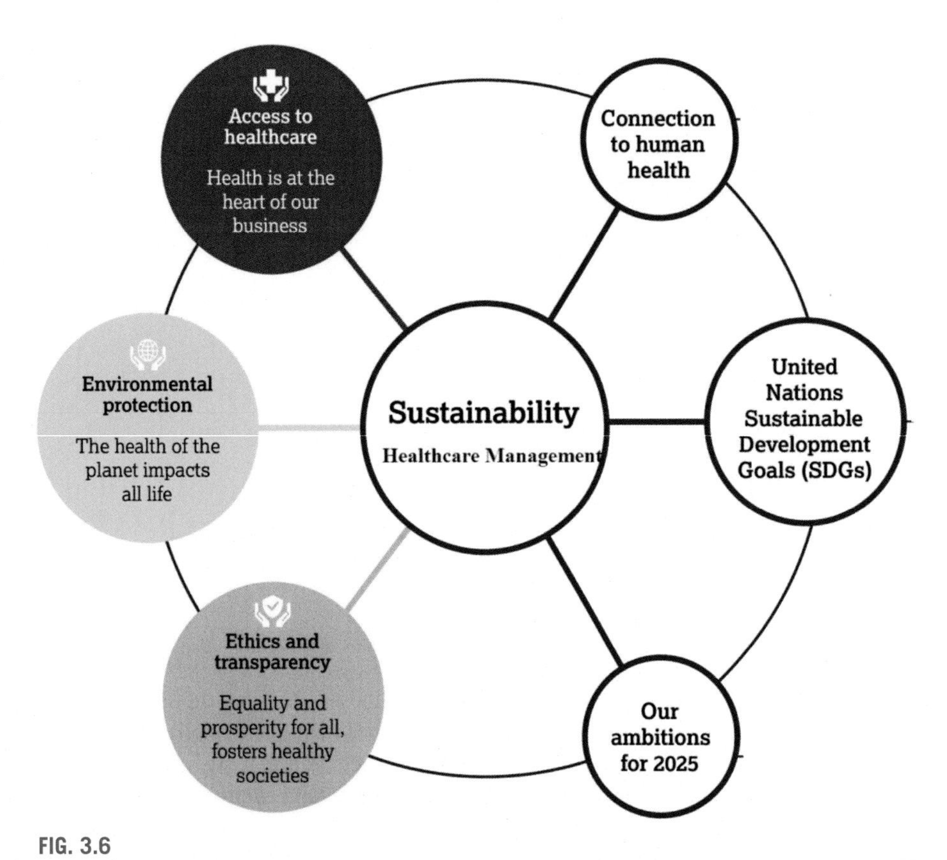

FIG. 3.6

Sustainable health-care management associated with different factors.

So, it is advisable that without compromising the quality of care for the patients, environmental sustainability is to be maintained. As human health is extremely linked with the quality of the environment, it is necessary to keep vigilance over these two issues, simultaneously.

3.3.1 What is sustainability?

Sustainability can be defined as: "the ability of something to continue overtime for the benefit of humanity" [12]. In this connection, environmental sustainability is referred to: the quality of causing little or no damage to the environment and, therefore, able to continue for a long time [12].

From a business point of view, sustainability means: maintenance of continuity in flow in economic, social, and environmental sustainability. The same can also be applicable to health-care sustainability with the inclusion of factors such as disaster preparedness, public health-care workforce capacities, and research capabilities.

3.3.2 Hospital waste management

Inadequate facilities in the hospital, waste management has been an alarming factor for sustainability in health-care services. Generally, hospital wastes resulted from rigorous infection control and include single-use clinical equipment, disposable contaminated linen, excessive packaging for medical items, and materials contaminated by patient fluids or contact. In addition, food wastes from patient meals, paper, cardboard, and plastic, and general wastes also contribute to medical wastes. But, waste management systems serve as an expensive practice, as well as contributing to indirect carbon emission. Clinical waste management involves incineration, autoclaving, or chemical disinfection, followed by placement in a landfill [13]. So, recycling of clinical medical wastes is helpful for increasing sustainability in health-care facilities and should be a mandatory practice in hospitals fixed by the government.

3.3.3 Infrastructure and planning

Another area in which environmental sustainability can be improved is in developing health-care infrastructure. Under the SDGs program as organized by the WHO, special emphasis is given to bring awareness to public and government to upgrade the surrounding natural environment and efforts to conserve local habitats; maximizing the use of natural light and ventilation in order to minimize electricity use; and the use of reflective material and solar energy system to generate electricity are some of the important factors directly responsible for increasing efficiency in sustainable health-care practice in hospitals [14].

3.4 Community diseases prevention

Following are few important tips for maintaining good health in developing a healthy community.

3.4.1 Frequent washing of hands

Keeping personal hygiene, like a daily bath and washing hands with soap and water is the most convenient and easy habit to prevent transmission of many deadly communicable diseases such as COVID-19 (Fig. 3.7).

Keeping personal hygiene, like taking a daily bath and handwashing with soap and water is the simplest, and one of the most effective ways to prevent the transmission of many communicable diseases (Fig. 3.8).

Hand washing is the single most effective way to prevent the possibility of contamination with infections. Generally, touching someone or any contaminated objective (such as lift doors, elevator supporters) one can also spread a certain "germ" (a general term for pathogens like COVID-19 viruses) by touching another person. One can, herself or himself, also catch germs when touching contaminated objects or surfaces and then touching their own face (mouth, eyes,

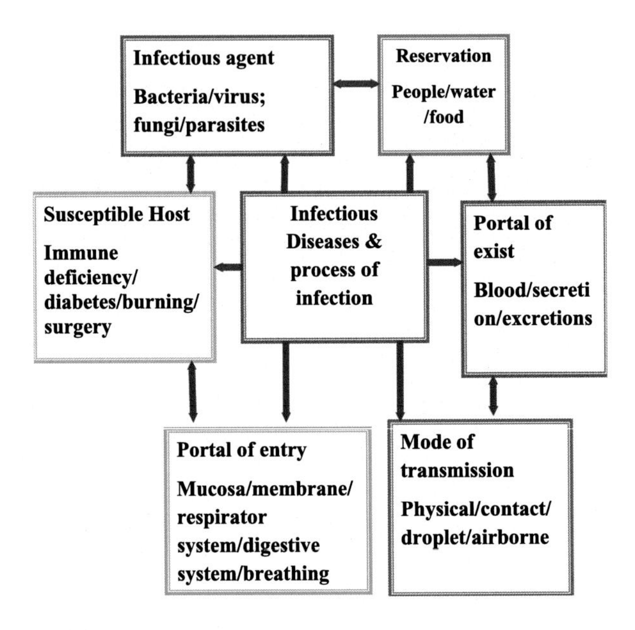

FIG. 3.7

Schematic presentation of various aspects of communicable diseases infection and prevention measures.

FIG. 3.8

Perfect process of hand washing.

nose). During a pandemic, it is also important to clean hands regularly, after one is free from public space or business. It is also advisable to wash the hands, immediately after using the washroom in order to reduce diseases transmission of gastrointestinal infection.

In order to maintain good overall health and wellness, it is necessary to maintain proper personal hygiene. Personal hygiene makes an individual healthy both in physical getup and mental condition and reduces the risk of diseases contamination.

3.4.2 Bath

By regular bathing, the dead cells of the skin are washed away followed by the regeneration of new skin cells. Regular bathing cleans the skin surface and helps in the problem of skin itching due to the accumulation of skin dead cells. Regular bath also helps from contamination of microbes and pathogens present in the surrounding environment.

3.4.3 Brushing teeth

An individual is advisable to brush teeth twice a day, once in the morning and once at night, to maintain good oral hygiene along with regular dental checkups.

3.4.4 Taking care of nails

Trimming of nails, twice a month, regularly reduces the possibility of risk of diseases contamination through the nail.

3.4.5 Wearing clean cloths

It is good hygiene practice to wear washed cloth after the bath of overall body cleaning. Dirty cloth harbor pathogen responsible for causing diseases.

3.4.6 Use of mask

When traveling outside, especially during communicable epidemic diseases like COVID-19, it is advisable to use the mask, and keep a social distance of about 5 ft. in order to save oneself from the possibility of contamination through droplets coming out in breathing or sneezing.

3.4.7 Restful sleep

Restful sleep is essential to well-being irrespective of age. It is recommended for all adults to sleep between 7 and 9 h a night.

3.4.8 Routine life

Routine is a word that can be applied for the perforation of day to day life in a sequential and regular manner. For example, brushing teeth immediately after getting from bed is routine. Even eating chips while watching TV or Netflix is a routine. To maintain good personal hygiene, it is necessary to have a routine lifestyle, which keeps an individual perfect foot with risk-free life.

3.4.9 Visit the doctor

Regular checking health with a home doctor or in hospital can help an individual in understanding the possible risk of infection, and making it easier to treat, timely.

3.4.10 Vaccination

Vaccination is a simple, safe, and effective way of protecting people against harmful diseases before they come into contact with them. It helps in developing immunity in the body to protect against infection like COVID-19. Most vaccines are given by injection, but some are given orally (by mouth) or sprayed into the nose. There are about 20 types of vaccines available for the most contagious diseases such as diphtheria, tetanus, pertussis, influenza, measles, and COVID-19. Together, these vaccines save the lives of up to 3 million people every year.

Currently, several vaccines are ready for taking prevention major from COVID-19. The first mass vaccination program was started in early December 2020. About 175.3 million vaccine doses have been administered. At least seven different vaccines (three platforms) have been administered. In December 2020, the WHO has issued an Emergency Use Listing (EULs) for the Pfizer COVID-19 vaccine (BNT 162b2). In addition, the WHO issued EULs for versions of the AstraZeneca/Oxford COVID-19 vaccine, manufactured by the Serum Institute of India and SKBio.

COVAXIN, India's indigenous (inactivated vaccine) *COVID-19 vaccine* by Bharat Biotech has been developed in collaboration with the Indian Council of Medical Research (ICMR)-National Institute of Virology (NIV).

COVAXIN®, India's indigenous ***COVID-19 Vaccine*** by Bharat Biotech is developed in collaboration with the Indian Council of Medical Research (ICMR)-National Institute of Virology (NIV). The vaccine is developed using *Whole-Virion Inactivated Vero Cell*-derived platform technology. Inactivated vaccines do not replicate and are therefore unlikely to revert and cause pathological effects. They contain dead viruses, incapable of infecting people but still able to instruct the immune system to mount a defensive reaction against an infection.

The inactivation technology for developing a vaccine has been a safe track record of >300 million doses of supplies to date. The same technology is adapted to manufacture vaccines against Seasonal Influenza, Polio, Pertussis, Rabies, and Japanese Encephalitis.

3.5 Noncommunicable diseases management

The main practices involve with noncommunicable (NCDs) chronic diseases include detecting, screening, and treating these diseases, and providing palliative care for people in need. NCDs are by far the leading cause of death in the world, representing 63% of all annual death. NCDs kill more than 36 million people each year. About 80% of deaths occur in low- and middle-income countries. The WHO is involved in the worldwide promotion of sustainable healthy life. For universal health coverage, the WHO has recommended the following:

- Focus on primary health care
- To maintain sustainable financing from state agency
- Availability of adequate access
- Availability of adequate workforce
- Support people's participation in national health policies
- Proper monitoring data bank

The WHO health target is mainly focusing on promoting health and well-being, and protecting health emergencies caused by natural disasters including biological pandemics [15]. Presently, having control over COVID-19 is in topmost priority for the WHO. Of the six WHO regions, the European region is the most affected by non-communicable diseases (NCD)-related morbidity and mortality [16]. Cardiovascular diseases, cancers, chronic respiratory diseases, and diabetes are among the leading causes of death and disability in the region [17]. In addition, overweight obesity has been one of the risk factors for NCDs. Prevention and control of NCDs are important during this pandemic. This is mainly due to the risk of getting infected with COVID-19 for NCDs likely of more chance [18,19]. So, it is advisable to observe lockdowns, social distancing, and restrictions in traveling as major prevention factors.

3.6 Community health assessment

The Affordable Care Act (ACA) or Obamacare came into existence on March 23, 2010. But, ACA's major provisions came into force in 2014. Since then, it was mandatory that the ACA requires all nonprofits hospitals to complete a Community Health Needs Assessment (CHNA) process every 3 years. A CHA is a systematic process involving the community to identify and analyze community health needs and assets, prioritize those needs, and then implement a plan to address significant unmet needs. Generally, completing the assessment is followed by hospital development strategies to address the significant community health needs identified in the CHA. The overall attempt for community health improvement is to develop a structure for addressing the determinants of health and illness in a community (Fig. 3.9).

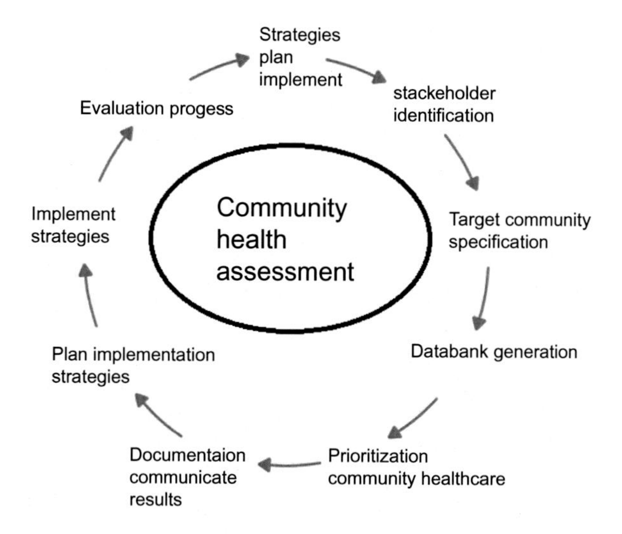

FIG. 3.9

Schematic representation of community health assessment process.

Community health assessment, planning, and implementation are crucial in governing "health for all" projects in a rural and remote locality. For community health development, it is necessary to identify the community health problem, and reshaping them for effective health services, even at the time of natural disaster or any other climatic change that will have adverse effects on health.

The key components associated with implementing health-care policies at the community level, irrespective of location, are as follows.

3.6.1 Internal partnerships

The community health improvement strategies can only be successfully implanted with the help of individuals and departments across the hospitals. In this regard, the support of hospital leadership plays a crucial role in securing funding and resources to implement desired strategies.

3.6.2 External partnerships

The inclusion of community stakeholders is an important issue in any health-care development program. It is better to have few new community stakeholders with some well-experienced stakeholders who can offer innovative insight and resources in bringing sustainability in health-care services. It is also necessary to associate the member of the population who can effectively put efforts to ensure plans are culturally appropriate. It has been noticed that many hospitals also take the help of local and state health departments in developing their CHAs. This sort of practice is helpful in the process of developing from the assessment phase to implementing strategies to address identified community health priorities. In addition, while going ahead in hospital leadership it is important to work out health services strategies on a priority basis.

3.6.3 Community assets

Availability of adequate assets can boost the practice of community health care. It is important to note that having few potential assets is more important than the availability of more number of nonfunctional assets. Examples of community assets may include emergency medical services, nursing or adult care home, mental health facilities, community health centers, health clinics, home health, and hospice care, school health services, medical and health transportation, and dental care providers.

3.6.4 Level of health care

It is necessary to understand the level of health care is given to patients in the hospital. As a result, the patients will understand their state of treatment and necessary requirements. The level of health care can be divided into four types: primary care, secondary care, tertiary care, and quaternary care.

Primary health care refers to the treatment to stop generalized symptoms and medical concerns related to the common cold, flu, or some pathogen infection. It may also be related to primary care for a broken bone, a sore muscle, a skin rash, or any other acute medical problems. Primary care providers (PCP) may be doctors, nurses, practitioners, or physician assistants. It is a regular practice that primary health-care provider refers the case to secondary health-care providers; mostly represent the specialized doctors, sometimes with the consent of patients. Once a patient is hospitalized and needs a higher level of specialty care within the hospital, he or she may be referred to tertiary care. Tertiary care requires highly specialized equipment and expertise. Quaternary care is considered to be an extension of tertiary care. It is super-specialized health care.

3.7 Child health-care development

Providing health-care life for children is fundamentally important in the society. Over the past several decades, the world has witnessed dramatic improvement in child health care and reducing the mortality rate of young children. As reported by the WHO, the number of children dying before the age of 5 was halved from 2000 to 2017, and more mothers and children are surviving today than before. All this could be possible due to the sincere efforts and involvement of international organizations such as the World Health Organization (WHO), UNICEF, the Partnership for Maternal (PMNCH), Newborn & Child Health, and International Rescue for Children living in a humanitarian setting. PMNCH is a multiconstituency partnership hosted by the WHO. The main target of PMNCH is to achieve universal access to comprehensive, high-quality reproductive maternal, newborn, and child health care. The International Children's Emergency Fund is a United Nations agency responsible for providing humanitarian and development aid to children worldwide. The agency is having members from 192 countries and territories. The main activities of the UNICEF include providing and promoting vaccination in order to enhance immunity potential in children, administering treatment for children and mothers with HIV, enhancing childhood and maternal nutrition, improving sanitation, promoting education (especially primary level), and providing emergency relief in response to disasters.

The WHO suggests that more than half of child death is due to conditions that can be easily prevented or treated by giving access to health care and improving their quality of life [3] 3 www.who.int › Newsroom › Fact sheet (2020) Children: improving survival and well-being.

The total number of under-5 deaths worldwide has reduced from 12.6 million in 1990 to 5.2 million in 2019. Since the last decade, the global under-5 mortality rate has dropped by 59%, from 93 deaths per 1000 live births in 1990 to 38 deaths in 2019. This is equivalent to 1 in 11 children dying before reaching the age of 5 in 1990, compared to the age of 1 in 27 children in 2019.

The child death rate under-5 age group is noted more in most of the developing or underdeveloped countries such as Africa. The highest number of mortality under-5 age group has been noted in Sub-Saharan Africa, with 1 in 13 children dying before his or her fifth birthday. Two regions, Sub-Saharan Africa and Central and Southern Asia, account for more than 80% of the 5.2 million under-5 deaths in 2019, while they only account for 52% of the global under-5 population. This is equivalent to half of under-5 deaths at global level, which occurred in five countries: Nigeria, India, Pakistan, the Democratic Republic of the Congo, and Ethiopia. Nigeria and India alone account for almost a third of all deaths. It has been noted that India, Nigeria, the Democratic Republic of the Congo, Pakistan, and China are more victimized countries in terms of child death rates for 5–9-year olds.

Infectious diseases are the most leading cause of death in children under-5 age group. These infectious diseases including diarrhea, malaria, pneumonia, preterm birth, birth asphyxia and trauma, congenital anomalies, and severe acute malnutrition remain the leading causes of death in rural and remote places of developing countries. The only way left out to save the children under-5 age group is to give breastfeedings and adequate nutrition, access to basic life-saving interventions such as skilled delivery at birth, postnatal care, timely vaccinations, and treatment for common childhood diseases.

In the case of old children, the main causes of health disorders are shifting of infectious early childhood diseases to later phase of teenage; injuries due to accidents (mostly caused by road traffic and drowning). So, it is necessary to take some prevention majors like improving the traffic rules and regulation and their strict implementation and carefully follow up of infectious diseases transferred from early teenage to late child adolescence. In addition, at the government and administrative level, care must be taken for primary education, sanitization, and development of remote communication, and the provision of adequate health-care providers.

In order to have reform in child health-care practices, the subclause 3 (3.2.1) of SDG is targeted to promote healthy lives and well-being for all children (Fig. 3.10).

The SDG Goal 3.2.1 is to end preventable deaths of newborns and under-5 children by 2030. There are two targets:

(i) Decrease newborn mortality to as low as 12 per 1000 live births in every country.

(ii) Decrease under-5 mortality to as low as 25 per 1000 live birth in every country.

Both targets 3.2.1 and 3.1.1 are closely linked with each other. The target 3.1.1 is to bring down under-5 child death to 70 deaths per 1,000,000 live births, whereas target 2.2.1 on ending all forms of malnutrition, as malnutrition, is a frequent cause of death for under-5 children. In 2029, 122 countries participated in SDG meeting to focus the discussion on under-5 mortality and it is expected that about 20 countries will meet the target by the end of 2030. It is expected that by meeting the target, the SDG could be able to reduce the number of under-5 deaths by 11 million between 2019 and 2030. Maximum efforts have been concentrated to prevent 80% of these deaths in Sub-Saharan Africa and southeast Asia.

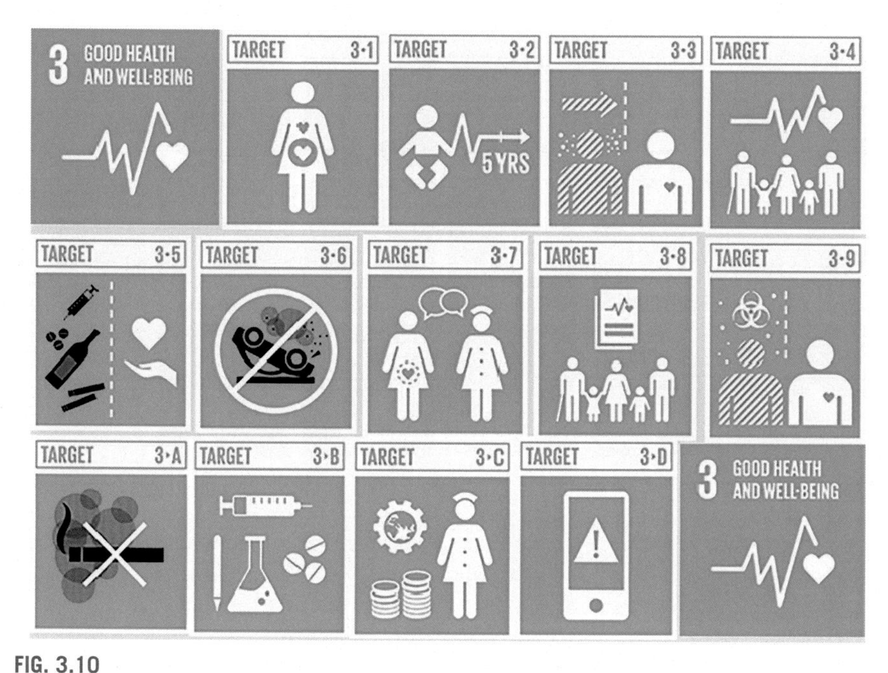

FIG. 3.10

Sustainable Development Goal-3.2.1 for health and well-being.

3.8 Rural health care and social inclusion

Still, the world faces disparities in health care between metropolitan and rural areas. Due to geographical location, lack of inadequate communication, poor economical condition, and political instability, rural people have been confronting better health services, for a long. Social inclusion in rural areas is often linked to health outcomes, but poor attention has been given by the government. So, it is necessary to understand the present scenario of rural health, and for the same purpose, we have to identify some indicators to understand the intensity of poor health-care services so as to take the rural health-care issues on priority basis.

Sustainable Development Goals, agenda for 2030 is consisting of 17 sustainable development goals with 169 challenges being supported by the United Nations as global efforts to reduce poverty, effect of extreme climatic changes, environmental pollution, and geographical inequalities which may ultimately lead to a healthy society with sustainability in overall redevelopment for well-being (Fig. 3.11).

This key strategic document is in continuation of the previous Eight Millennium Development Goals (started in 2000) committed to combating poverty, hunger, diseases, and illiteracy, promoting gender equality, and ensuring environmental sustainability until 2015.

Out of 17 SDGs, SDG 3 is closely related to “ensure healthy lives and promote well-being for all at all ages.” The SDG declaration emphasizes that to achieve the

FIG. 3.11

Sustainable development goals agenda supported by the United Nations.

overall health goal, "we must achieve universal health coverage (UHC) and access to quality health care." SDG3 is targeted to ensure health and well-being for all, at every stage of life, irrespective of any geographical location, also.

3.8.1 Social inclusion and health care

Social inclusion referred to gather a group of people irrespective of caste and community to make them feel to take part in society. People take part in society through markets (e.g., labor, or credit), services (access to health, education), and spaces (e.g., political, physical). Where individuals or groups of individuals are excluded, or feel on the margins of society, there is often a direct impact on their health. Social inclusion can counteract isolation and increase community participation, which helps to alleviate health problems, especially mental health issues such as anxiety and depression.

Now, people have started thinking to develop policies and the delivery of services to promote social inclusion for people suffering from chronic illness (physical or mental) or disability. The chronic health issues may be restricted as to their personal problem, but essential issues from the community point of view. This sort of problem is common in many of the developing countries, especially in rural localities.

For the first time, the world summit for social development was held in March 1995 with the target to develop the concept on social integration to create an inclusive society, "a society for all," as one of the key goals of social development. The Copenhagen Declaration and Programme of Action resulted as key outcomes of the summit, pledged to eradicate poverty, create full employment, and promote social integration overriding objective of development. Member states made a commitment to promote social integration through fostering inclusive societies that are stable, safe, and respective diversity, equality of opportunity, and participation of all people, including disadvantaged and vulnerable groups and persons. In 2000, the Millennium Declaration Goals included social integration as a main theme of peace, security, development, and human right for "health for all." The process of social integration also serves an important part in achieving the goal of sustainable development, particularly in view of the direct and indirect impacts to be brought by ongoing climate change and food insecurity on the most vulnerable population.

3.8.2 Social exclusion impact on health care

Social exclusion is defined as social disadvantages and lack of resources, opportunity, participation, and skill. Social exclusion through social discrimination can be responsible for psychological damage and harm to health through long-term stress and anxiety. Social exclusion has a direct impact on the health system, which is ultimately responsible for economic declination and the creation of social inequality that directly influences health. The WHO reports say that social inequalities and discrimination are mainly responsible for adverse health outcomes and life expectancy across groups in society.

Social exclusion and poverty pose a major threat to the health and well-being of millions of people belonging to rural areas of developing countries. In order to overcome such problems, and achieving universal coverage in health care to ensure a lot of easily affordable access have been projected by the government or private agencies. Low- and middle-income countries have introduced a variety of health financing reforms, which aim to introduce prepayment for health care at affordable prices for low socioeconomic groups as well as targeted subsidies for other vulnerable groups. Examples of this social health protection (SHP) schemes include community-based health insurance in India and Burkina Faso, vouchers and conditional cash transfers for maternal health in India and Pakistan, and national and social health insurance in Ghana and Vietnam. In spite of the implementation of such a scheme, still, more than one billion people worldwide are lacking affordable health care.

The World Health Organization (WHO) says social exclusion (SE) is one of the major causes for discrimination in the community [20–22]. This is the main reason why people belonging from rural regions mostly are incapable of participating fully in all sort of social activities [23], while the group of people under SI enjoy all sort of economic, social, cultural, and health-care benefits [24]. The complexity of poor health-care services due to SE is quite diversified in nature like deprivations in areas such as social relations, material resources, access to health services, and housing infrastructure [25,26]. SE is also responsible for generating inadequate labor availability, good nutrition deficit, and nonavailability of health-care delivery professions. These factors may directly impact health and well-being [27–29].

3.9 Water security for public health care

The rural people are more susceptible to ill health than the urbanized locality. This is mainly due to an inadequate supply of pure and drinkable water. For public health purposes, fresh and readily available water is needed for drinking, domestic use, food production, or recreational purposes.

Table 3.1 provides the selection parameters for drinking water as approved by the WHO.

In 2010, the UN General Assembly exclusively recognized that every human has the right to have sufficient, safe, and secure quality water continuously available for human use in domestic and as drinking water.

Sustainable Development Goals (SDGs) target 6.1 emphasis on the availability of safe and drinking water for public use at an affordable cost. In addition, targets 6.3, 6.5, and 6.6 are also focused on water quality, water storage, and safety, respectively (Fig. 3.12).

3.9.1 Poor water quality and health

Poor water quality and lack of sanitation are responsible for causing contagious diseases such as diarrhea, dysentery, hepatitis A, typhoid, and many other waterborne diseases. Inappropriate management of water, improper water storage systems, and lack of city water filter systems expose individuals to preventable health risks. Inadequate management of urban, industrial, and agricultural wastewater has maximum possibility of causing health disorders, especially for rural people.

Diarrhea is the most widely known disease linked to contaminated food and water. In 2017, over 220 million people required preventive treatment for schistosomiasis—an acute and chronic diseases caused by parasitic worms contracted through exposure to infected water. In many parts of the world, transmitted diseases caused by insect vectors are waterborne diseases due to improper storage of water in cooler or other purposes.

3.9.2 Preventive measures for water burn diseases

Mostly, food poisoning is caused by eating or drinking food or beverages contaminated by bacteria, parasites, or viruses. These organisms are passed in the faces of animals and infected people. Generally, stomach cramps like gastrointestinal symptoms are caused by food and waterborne illnesses. One can also get sick from swimming in contaminated water or from close contact with someone else who is ill. Certain groups of people develop serious complications. These include young children, pregnant women, senior citizens, and persons having low immune potential.

There are many ways to prevent this illness:

(i) Flush before and after using the toilet.
(ii) Use detergent while washing the floor.
(iii) It is recommended to use a chlorine-based disinfectant.
(iv) Carefully and frequently handwashing with standard soap.
(v) Wash hands before and after preparing food or eating.
(vi) Wash and peel vegetables before cooking.
(vii) Use pasteurized dairy products such as milk, cheese, yogurt, and ice cream.
(viii) Use boiled meat (71 °C/160 °F) before cooking.
(ix) Wash hands, kitchen work surfaces, and utensils after contact with raw meat or poultry.
(x) Clean and sanitize all utensils, equipment, and surfaces (cutting boards, work counters) before and after each use. Be sure to sanitize food contact surfaces with a sanitizing solution.

Table 3.1 Drinking water quality as prescribed by WHO.

Parameter	World Health Organization	EU	US	China	Canada	India
1,2-Dichloroethane		3.0 μg/L	5 μg/L	-do-	-do-	
Acrylamide		0.10 μg/L	TT[a]	-do-	-do-	
Aluminum		0,2 mg/L		-do-		0.03 mg/L
Antimony	Ns	5.0 μg/L	6.0 μg/L	-do-	-do-	
Arsenic	10 μg/L	10 μg/L	10 μg/L	50 μg/L	50 μg/L	0.05 mg/L
Barium	700 μg/L	Ns	2 mg/L	-do-	-do-	
Benzene	10 μg/L	1.0 μg/L	5 μg/L	-do-	-do-	
Benzo(*a*)pyrene	-do-	0.010 μg/L	0.2 μg/L	0.0028 μg/L	0.0028 μg/L	
Beryllium						
Boron	2.4 mg/L	1.0 mg/L		-do-	5.00 mg/L	1.0 mg/L
Bromate	-do-	10 μg/L	10 μg/L	-do-	-do-	0.01 mg/L
Cadmium	3 μg/L		5 μg/L	5 μg/L	5.00 mg/L	0.01 mg/L
Calcium		5 μg/L			200 mg/L	75 mg/L
Chromium	50 μg/L	50 μg/L	0.1 mg/L	50 μg/L (Cr6)	0.050 mg/L	0.05 mg/L
Cobalt					-do-	
Copper	-do-	2.0 mg/L	1.3 mg/L[b]	1 mg/L	1.00 mg/L	0.05 mg/L
Cyanide	-do-	50 μg/L	0.2 mg/L	50 μg/L	-do-	0.05 mg/L
Epichlorohydrin	-do-	0.10 μg/L	TT[a]	-do-	-do-	
Fluoride	1.5 mg/L	1.5 mg/L	4 mg/L	1 mg/L	-do-	1.0 mg/L
GoLd					no limit listed	
Hardness					0–75 mg/L = soft	300 mg/L
Iron		0.2 mg/L			0.300 mg/L	0.3 mg/L
Lanthanum					No limit listed	
Lead		10 μg/L	15 μg/L[b]	10 μg/L	10.0 μg/L	0.05 mg/L
Magnesium					50.0 mg/L	30 mg/L

Continued

Table 3.1 Drinking water quality as prescribed by WHO.—cont'd

Parameter	World Health Organization	EU	US	China	Canada	India
Manganese		0. 05 mg/L			0.050 mg/L	0.1 mg/L
Mercury	6 μg/L	1 μg/L	2 μg/L	0.05 μg/L	1.00 μg/L	0.001 mg/L
Molybdenum					no limit listed	
Nickel		20 μg/L		-do-	-do-	
Nitrate	50 mg/L	50 mg/L	10 mg/L (as N)	10 mg/L (as N)	-do-	45 mg/L
Nitrite	3 mg/L	0.50 mg/L	1 mg/L (as N)	-do-	-do-	
Pesticides—Total	-do-	0.50 μg/L	-do-	-do-	-do-	Absent
Pesticides (individual)	-do-	0.10 μg/L	-do-	-do-	-do-	
pH					6.5–8.5	6.5–8.5
Phosphorus					no limit listed	
Polycyclic aromatic hydrocarbons		0.10 μg/L			-do-	
Potassium					no limit listed	
Scandium					no limit listed	
Selenium	40 μg/L	10 μg/L	50 μg/L	10 μg/L	10.0 μg/L	0.01 mg/L
Silicon					no limit listed	
Silver					0.050 mg/L	
Sodium					200 mg/L	
Strontium					No limit listed	
Tetrachloroethene and trichloroethene	40 μg/L	10 μg/L	10 μg/L		-do-	
Tin					No limit listed	

Titanium					No limit listed	
Tungsten					No limit listed	
Uranium					0.10 mg/L	
Vanadium					No limit listed	
Zinc					5.00 mg/L	5.0 mg/L
vinyl chloride		0,50 μg/L				
Chlorides		250 mg/L				250 mg/L
electrical conductivity		2500 μS cm^{-1} at 20 °C				
Total dissolved solids						500 mg/L
Sulfate						200 mg/L

[a] *TT (treatment technique). The public water system must certify that the combination of dose and monomer level does not exceed: Acrylamide = 0.05% dosed at 1 mg/L (or equivalent); Epichlorohydrin = 0.01% dosed at 20 mg/L (or equivalent) [11].*

[b] *Action level; not a concentration standard. A public water system exceeding the action level must implement "treatment techniques" which are enforceable procedures [13].*

Indicates that no standard has been identified by editors of this article and ns indicates that no standard exists. μg/L → micro grams per liter or 0.001 ppm, mg/L → 1 ppm or 1000 μg/L.

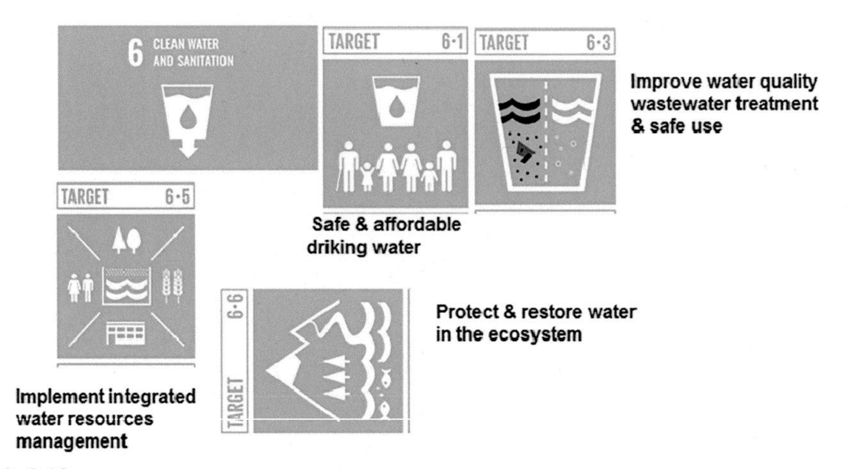

FIG. 3.12

Sustainable Development Goal 6 and other related goals for clear water and sanitization.

3.9.3 Drinking water services

Generally, safe and secure water comes from well-managed resources or artificial-made community water filter systems located on premises, available when needed, and free from fecal and priority chemical contamination. So, it is necessary to be alert while using safe water without any waste. SDGs target 6.1 mainly focus on universal and equitable access to safe and affordable drinking water.

As the WHO reports, in 2017, 71% of the global population (5.3 billion people) used a safely managed drinking water service located on premises, available when needed, and free from contamination. About 785 million people still lacked even a basic drinking water service, defined as drinking water from an improved source, provided collection time is not more than 30 min for a round trip. This includes 144 million people who collect untreated surface water for drinking.

3.9.3.1 WHO's efforts

As an international authority on public health, the WHO puts maximum efforts to prevent transmission of waterborne diseases, and advising the government on the development of health-based targets and regulations. The WHO has published a series of water quality guidelines, water contamination, safe use of wastewater, and safe recreational water environments [30–37].

The WHO's water quality guidelines are based on risk management. Since 2004, the guideline for drinking-water quality has been promoting the framework for safe drinking water. The WHO also provides quality guidelines to the member of states, from time to time. This is mainly to supply updated information under variable climatic conditions and natural disasters. Since 2014, the WHO has been testing household water treatment products based on guidelines to evaluate household water treatment technologies. This sort of testing practices is mainly to strengthen policy, regulatory, and monitoring mechanisms at the national level to support appropriate targeting and consistent and correct use of such products.

The WHO also works jointly with UNICEF in the areas related to water, sanitation, and hygiene in health-care facilities. In 2015, both the agencies, jointly sponsored a project on "WASH FIT" (Water and Sanitation for Health Facility Improvement Tool) to guide small, primary health-care facilities in low- and middle-income settings through a continuous cycle of improvement through assessments, prioritization of risk, and definition of specific, targeted action (Fig. 3.13).

In this regard, the WHO reported that in 2019, the virtual steps were taken at the country level to improve water quality, sanitation, and hygiene in health-care facilities.

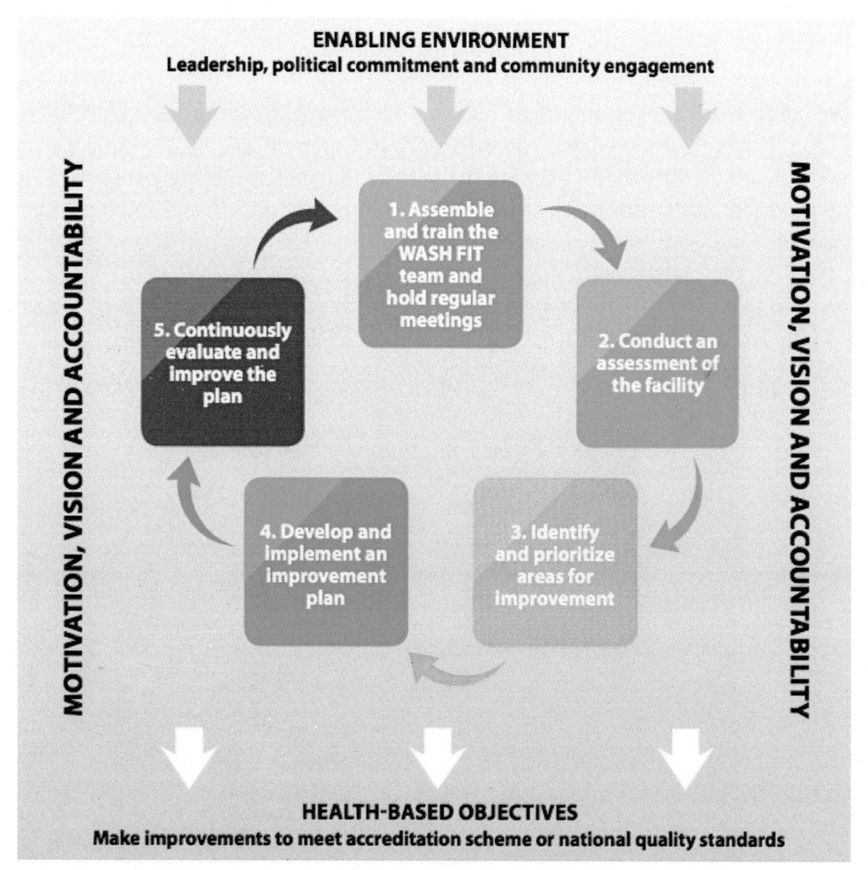

FIG. 3.13

Operational protocol of "WAH FIT" protocol and guidelines being sponsored by the WHO and UNICEF, jointly.

"WASH FIT" project is mainly focused on four broad areas: water, sanitation (including health-care waste management), hygiene (hand hygiene and environmental disinfection), and management (Fig. 3.14). For a better workout of the project, each area consists of a list of indicators and targets for achieving minimum standards for maintaining a safe and clean environment.

WASH FIT is a challengeable and highly inspiring project for improving and giving value-added touch to the project related to water management, sanitation and hygiene, and health-care waste management infrastructure and services in health-care facilities in low- and middle-income countries (LMIC).

The WASH FIT protocol explains how to follow the sequential process for assessing and improving health-care services and related matters in the most affordable and dedicated ways. It is basically well framed in accord with the WHO guidelines for drinking-water quality (WHO 2011) [38] and goes beyond water safety including sanitation and hygiene, health-care waste, management, and staff empowerment.

WASH FIT project boost in improving health-care facilities in order to reduce maternal and new born mortality and improves the quality of care maximum care and precaution will be taken in a dignified way for managing the process of delivering child. It has been observed that countries such as Cambodia, Chad, Ethiopia, Liberia, and Mali are on the way to adapting the WASH FIT project. In this connection, it is expected that at the government level well plan policy design and implementation

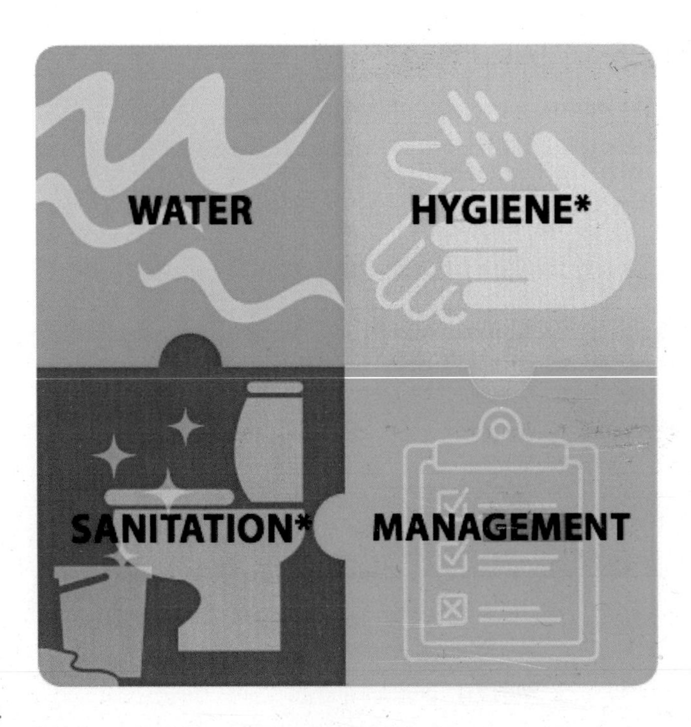

FIG. 3.14

Four WASH FIT coverage jointly sponsored by the WHO and UNICEF.

should be undertaken to develop WASH infrastructure. Now, it is high time to plan and go for multisectoral collaborations before undertaking the WASH FIT project for a community health-care development.

3.10 Food security and health care

The year 2019–21 has witnessed the pandemic scenario of COVI-19 and the problem of undernourishment at the global level. Priority to the COVID-19 pandemic, about 820 million people did not have enough to eat in 2018, up from 811 million in the previous year, which is the third year of increase in a row. These facts and figures have given the immense challenge of achieving the Sustainable Development Goal of Zero Hunger by 2030. This is the main theme of the report "The State of Food Security and Nutrition in the World" released in 2019 [39].

The world is not on track to achieve Zero Hunger by 2030, and if the present trend continues, the number of people affected by hunger will surpass 840 million. Since 2015, the overall number for hunger is in the steady increasing pattern. According to the World Food Program, 135 million suffer from acute hunger largely due to man-made conflicts, climate change, and economic downturns. The COVID-19 pandemic could able to increase hunger enormously by the end of 2020. In order to nourish the more than 690 million people who are the hunger today, a profound change of the global food and agriculture system is needed. This is mainly due to the additional two billion people the world will have by 2050.

As stated earlier, in 2015 the global community adopted the 17 Global Goals for Sustainable Development (Fig. 3.15) to upgrade the standard of life from well-being

FIG. 3.15

Sustainable Development Goal-2 and other related goals showing to eradicate hunger by 2030.

angle. Goal 2-Zero hunger—pledges to eradicate hunger and achieve food security, improve nutrition quality, and promote sustainable agriculture and is the priority of the World Food Program.

Five years passed, since the adaption of SDGs, the 2020 report notes that progress had been made in some areas, such as improving maternal and child health, expanding access to electricity, and increasing women's representation in government.

References

[1] Paim J, Travassos C, Almeida C, Bahia L, Macinko J. The Brazilian health system: history, advances, and challenges. Lancet 2011;377(9779):1778–97.
[2] Victora CG, Barreto ML, do Carmo Leal M, Monteiro CA, Schmidt MI, Paim J, et al. Health conditions and health-policy innovations in Brazil: the way forward. Lancet 2011;377(9782):2042–53.
[3] World Health Organization. International statistical classification of diseases and related health problems: Tabular list. Vol. 1. World Health Organization; 2004.
[4] Ahmad OB, Boschi-Pinto C, Lopez AD, Murray CJ, Lozano R, Inoue M. Age standardization of rates: a new WHO standard. vol. 9. Geneva: World Health Organization; 2001 [10].
[5] Costa AM. Social participation in the achievement of health policies to women in Brazil. Cien Saude Colet 2009;14:1073–83.
[6] Colpani V, Baena CP, Jaspers L, et al. Lifestyle factors, cardiovascular disease and all-cause mortality in middle-aged and elderly women: a systematic review and meta-analysis. Eur J Epidemiol 2018;33:831–45.
[7] Rechel B, McKee M. Facets of public health in Europe. European observatory on health systems and policies series. Maidenhead, UK: Open University Press; 2014.
[8] United Nations Millennium Development Goals. Eradicate extreme poverty & hunger; 2008. http://www.un.org. >milleniumgoals>poverty.
[9] Mdg Monitor. Tracking the millennium development goals; 16 May 2011. http://mdgs.un.orgs. >product>progress2011.
[10] List of goals, targets, and indicators [PDF] Siteresources.worldbank.org.
[11] World Health Organisation Regional Office for Europe. Environmentally sustainable health systems: a strategic document [internet]. Copenhagen: WHO Europe; 2017. http://www.euro.who.int/__data/assets/pdf_file/0004/341239/ESHS_Revised_WHO_web.pdf?ua=1.
[12] Cambridge Dictionary. Sustainability. [Internet] [Updated 2019; cited 2019 Jan 22]; Available from: https://dictionary.cambridge.org/dictionary/english/sustainability; 2019.
[13] Blue Environment Pty Ltd. Hazardous waste in Australia 2017 [internet]. Australian Government Department of Environment and Energy: Canberra; 2017.
[14] Global Green and Healthy Hospitals. Buildings. [Internet] [updated unknown; cited 2019 Jan 18]; Available from: https://www.greenhospitals.net/buildings/; 2015.
[15] Kluge H. A new vision for WHO's European region: united action for better health. Lancet Public Health 2020;5:e133–4.
[16] WHO. Noncommunicable diseases country profiles. 2018. Geneva: World Health Organization; 2018.
[17] WHO. Overweight. European Health Information Gateway; 2020. https://gateway.euro.who.int/en/indicators/h2020_6-overweight/visualizations/#id=17077.

[18] Wang B, Li R, Lu Z, Huang Y. Does comorbidity increase the risk of patients with COVID-19: evidence from meta-analysis. Aging 2020;12:6049–57.
[19] WHO. Noncommunicable diseases in emergencies. Geneva: World Health Organization; 2016.
[20] Popay J, Escorel S, Hernández M, et al. Understanding and tackling social exclusion. In: Final Report to the WHO Commission on Social Determinants of Health from the Social Exclusion Knowledge Network. Geneva: World Health Organization; 2008.
[21] Mathieson J, Popay J, Enoch E, et al. Social exclusion. Meaning, measurement and experience and links to health inequalities. A review of literature. Geneva: WHO Social Exclusion Knowledge Network; 2008.
[22] WHO. Poverty, social exclusion and health systems in the WHO European region. Copenhagen: WHO Regional Office for Europe; 2010.
[23] Millar J. Social exclusion and social policy research: defining exclusion. In: Abrams D, Christian J, Gordon D, editors. Multidisciplinary handbook of social exclusion research. Chichester: John Wiley & Sons Ltd; 2007. p. 1–16.
[24] Standing Senate Committee on Social Affairs Science and Technology. In from the Margins, Part II: Reducing Barriers to Social Inclusion and Social Cohesion. Ottawa, Ontario, Canada: Senate; 2013.
[25] Courtin E, Knapp M. Social isolation, loneliness and health in old age: a scoping review. Health Soc Care Community 2017;25:799–812.
[26] Gomez SL, Shariff-Marco S, DeRouen M, et al. The impact of neighborhood social and built environment factors across the cancer continuum: current research, methodological considerations, and future directions. Cancer 2015;121:2314.
[27] Marmot M. Status syndrome: how your social standing directly affects your health. London: Bloomsbury Publishing Plc; 2004.
[28] Wilkinson R, Pickett K. The spirit level: why more equal societies almost always do better. London: Penguin Books Ltd; 2010.
[29] Sen A. Social exclusion: concept, application, and scrutiny. Manila, Philippines: Asian Development Bank; 2000.
[30] WHO Guidelines for drinking-water quality. 3rd ed. Incorporating the first and second addenda, vol. 1.
[31] WHO. Heptachlor and heptachlor epoxide in drinking-water. In: Background document for preparation of WHO Guidelines for drinking-water quality. Geneva: World Health Organization; 2003 [WHO/SDE/WSH/03.04/99].
[32] WHO. Uranium in drinking-water. In: Background document for preparation of WHO Guidelines for drinking-water quality. Geneva: World Health Organization; 2003 [WHO/SDE/WSH/03.04/118].
[33] WHO. Xylenes in drinking-water. In: Background document for preparation of WHO Guidelines for drinking-water quality. Geneva: World Health Organization; 2003 [WHO/SDE/WSH/03.04/25].
[34] WHO. Zinc in drinking-water. In: Background document for preparation of WHO Guidelines for drinking-water quality. Geneva: World Health Organization; 2003 [WHO/SDE/WSH/03.04/17].
[35] WHO. Diflubenzuron in drinking-water: use for vector control in drinking water sources and containers. In: Background document for preparation of WHO Guidelines for drinking-water quality. Geneva: World Health Organization; 2008 [WHO/HSE/AMR/08.03/6].

[36] WHO. Novaluron in drinking-water: use for vector control in drinking water sources and containers. In: Background document for preparation of WHO Guidelines for drinking-water quality. Geneva: World Health Organization; 2008 [WHO/HSE/AMR/08.03/11].

[37] WHO. Pirimiphos-methyl in drinking-water: Use for vector control in drinking-water sources and containers. In: Background document for preparation of WHO Guidelines for drinking-water quality. Geneva: World Health Organization; 2008 [WHO/HSE/AMR/08.03/15].

[38] WHO. Guidelines for drinking-water quality. 4th ed. WA 675: NLM classification; 2011. ISB 97892 41548151.

[39] Food and Agriculture Organization of the United Nations. The state of food security and nutrition in the world, safeguard against economic slowdowns and downturns. Rome: Food and Agriculture Organization of the United Nations; 2019. ISBN 978-92-5-1315-0-5.

CHAPTER

Health-care information technology and rural community

4

4.1 Introduction: Health information technology

Health information technology (HIT) refers to a set of comprehensive and well-planned data bank that can be accessed as immediate reference about health care by physicians to assist in the state of health condition of a patient for better treatment and cure (Fig. 4.1).

The HIT can be defined as: "*the application of information processing involving both computer hardware and software that deals with the storage, retrieval, sharing, and use of health-care information, data, and knowledge for communication and decision making*" [1].

It could be private as well as public clinics, hospitals, and doctor's private chamber. It also collects, organizes, and manages patient's electronic medical records. The use of health-care software can be helpful in improving the quality of patient treatment. Besides this, it also takes care of administrative aspects and patent's medical data. In addition, HIT provides numerous opportunities for improving and transforming health care, which include: reducing human errors, improving clinical outcomes, facilitating care coordination, improving practice efficiencies, and tracking data over time [2–4].

4.2 Benefits of health information technology

The benefits of health information technology (HIT) include its ability to store and retrieve data; ability to quickly communicate patient information in a legible format; and improved medication safety through increased legibility, which potentially decreases the risk of medication errors. Since the last two decades, transformation of HIT has been reshaping the procedural application of health services in both urban and rural areas. Meanwhile, numerous technology tools are becoming available to improve health for all. It has been noted that most consumers have been encountering the benefits of HIT through an electronic health record, or EHR, available in hospitals or private clinic.

Healthcare Strategies and Planning for Social Inclusion and Development. https://doi.org/10.1016/B978-0-323-90446-9.00004-6

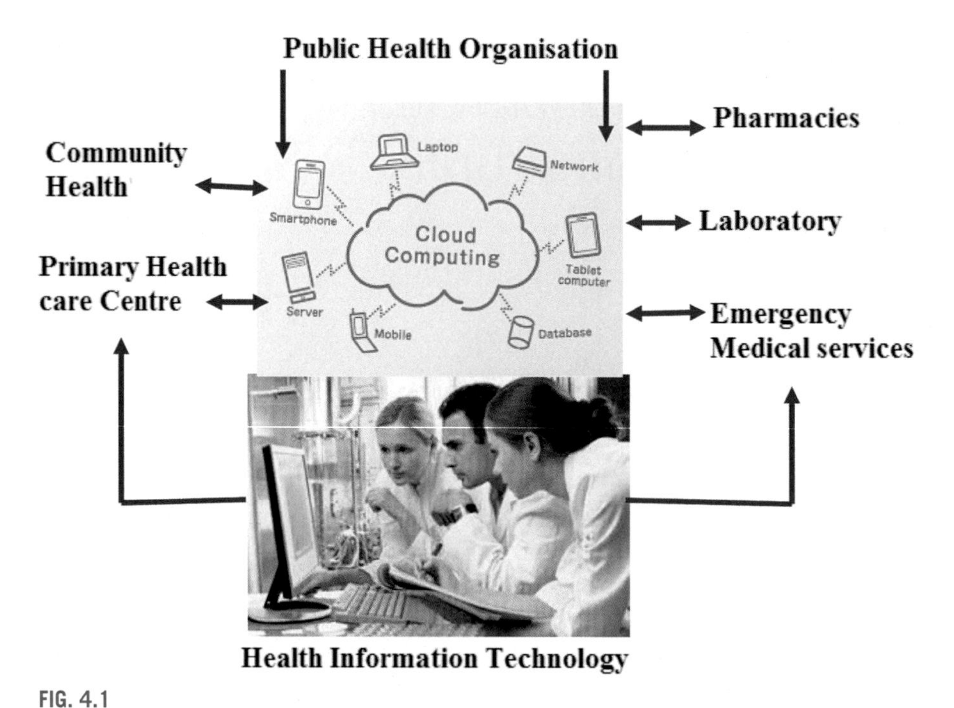

FIG. 4.1

How health information integrated with different services is related to health.

Health information technology (HIT) plays a significant role in improving the quality, safety, effectiveness, and delivery of health-care services in rural communities (Fig. 4.2).

It has been a challenging issue to provide updates on health information and monitor information technology in developing countries, including the rural peripheral communities. HIT plays a significant role in storing, securing, and transferring protected health information electronically within health-care systems and community setting. Key components of HIT are as follows.

Electronic health record (EHR)

An electronic health record (EHR) is an electronic version of patient's medical history that is maintained by the provider over time, and may include the key administrative clinical data relevant to that person's care under a particular provider, including demographics, progress notes, problems, medication, and other relevant data on health care.

Electronic transmittal of medical test results

It is a short of highly secure digital network to deliver up-to-date records whenever and wherever the patient or clinician needs them. It has highly configurable application which receives, process, and stores information generated by the medical laboratory process.

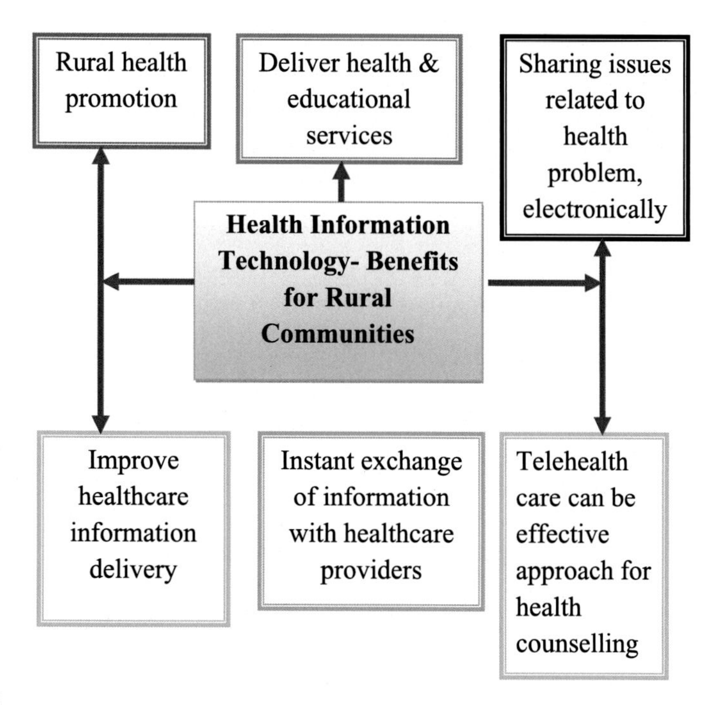

FIG. 4.2

Benefit of health information technology for rural people.

Telehealth

Telehealth is the use of digital information and communication technologies, such as computer and mobile devices, to access health-care services and help in remotely managing health care. More on telehealth is discussed in a separate section, in this chapter.

Telemonitoring applications

Telemonitoring is referred to the use of information technology to monitor patients at a distance. It is most promising for monitoring chronic illness such as cardiopulmonary disease, asthma, and heart failure in the home.

Patient health portal

A patient health portal is a secure online website that gives patients convenient 24-h access to personal health information from anywhere with an Internet connection using a secure password to receive information and also do communication on personal health issues.

Electronic prescribing

It is referred to the practice of safety and quality of prescribing process related to patient's need. E-prescribing allows providers in the ambulatory care setting to send prescriptions electronically to the pharmacy and can be a stand-alone system or part of an integrated electronic health record system.

Applications of health information technology in rural communities support instant communication and connection along with devices and machines that reduce the reliance on city- or urban-based facilities. In rural health care, this has been great challenge, especially for patients receiving ongoing or complex treatment for chronic issues, who may need to travel long distances for various health-care services. HIT provides lot of benefits for rural people by increasing effectiveness of a health-care institute. It ensures the accuracy of data collected and thus reduces any chance of medical error. Besides this, HIT saves time and effort, which is spent daily on managerial tasks by streamlining operations, so that organizations can concentrate on patient treatment and follow-up.

4.2.1 What HIT profession do?

Health information technology (HIT) experts handle the technical aspects of managing patient health information. Such alerts in the electronic health record (EHR) are a standard mechanism for the use of HIT for the prevention of potential missed quality and patient safety events. For example, immunization alerts have led to 12% increase in well-child and a 22% increase in sick-child immunization administration [5], and drug alerts have been associated with a 22% decrease in medication prescription errors [6].

On the basis of position, the IT professionals are responsible for the implementation and management of electronic health records (EHRs) and other systems that store patient-related data. The quality health care mainly depends on the efficacy and working style of health professionals. Generally, the health-care teams to drive improved outcomes, lowered costs, and new developments in patient care. Their roles, mainly, focus on patient medical records such as health information technician, health information manager, and health information specialist. In addition, the HIT professionals take care of traditional information technology works, such as the design, implementation, and maintenance of the computer networks. Generally, medical documents contain confidential patient information, so it is necessary for HIT professional to maintain patient privacy and comply with strict government regulation.

4.2.2 Importance of HIT

The advantages of health information technology are that it facilitates communication between health workforce and people from rural areas, improves health-care quality, tracks health status, and ensures medication safety (Fig. 4.3).

HIT plays a significant role in the health-care system in improving quality services. It helps to coordinate all existing medical knowledge about a patient and to incorporate new information on a real-time basis. It is vitally important in the digital age to maintain transparency, accessibility, and personalization of health-care-related issues.

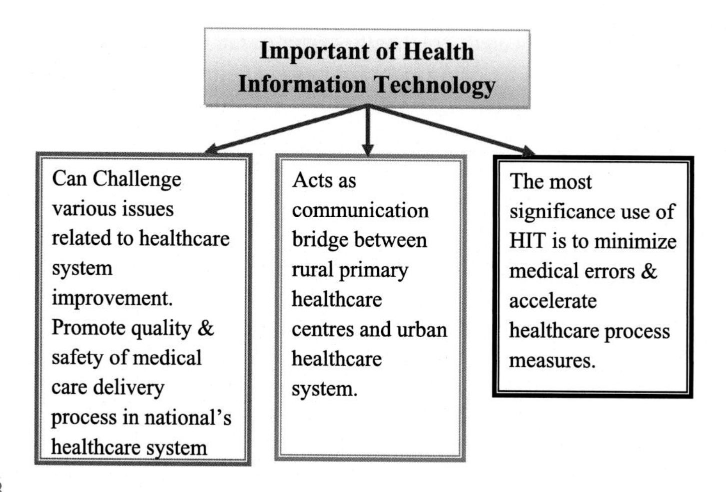

FIG. 4.3

Importance of health-care information technology.

Health information technology can bring remarkable changes in quality, safety, effectiveness, and delivery of health-care services in rural communities. HIT can work as a communication bridge between rural communities and urban areas for delivering quality, safety, and effective health-care services. HIT can meet the challenges for the development of rural facilities related to health services. The key components of HIT include electronic health record (EHR), secure digital networks, electronic transmittal of medical test results, telehealth, and electronic prescribing.

4.2.3 Future prospect of HIT

The last two decades have witnessed the how information technology has challenged the efficacy of health services. Information technology has completely changed the scenario of health services by making it accessible to people, increasing its efficiency and reducing a lot of costs associated with health-care management system. It is expected that health-care information technology will reshape the rural health services and bring it to a sustainable pattern. Following are the few important aspects of HIT which can reshape the health-care services in due time.

Better accessibility

Health-care information technology is going to be an important tool for reshaping health-care process for the people living in rural and remote areas. Due to inadequate infrastructure and poor health-care service in rural communities, residents often have to spend time and money traveling long distances to receive special treatment in city hospitals. In this connection an electronic health record (EHR) is crucial to rural population through which a health-care professional can view patient's medical history no matter where they are located. Telehealth systems can make remote consultations and diagnosis easier and more convenient for both patients and clinicians.

Increasing efficiency

By the application and practice of HIT, the physicians can not only save time but also increase their efficiency in working style. It also gives the physicians an opportunity to serve more number of people within limited scheduled time. Besides this, HIT benefits can be obtained when problem of shortfall of staff arises, in case of emergency services. So, federal and state policy makers, private payers, and systems leaders are interested in implementing operation status of HIT in order to reduce unnecessary burden and increase the efficiency of health-care delivery system, and allocate resources in order to improve value in health care.

Electronic health records

Electronic health records (EHRs) are sometimes known as electronic medical records in place of the paper copy, according to HIT. The objective of the EHR is to provide medical professionals health information through a database that allow certain people to access health information even if the immediate doctor's office is closed. Sometimes a paper prescription can be easily lost or even misread if the handwriting is not legible. E-prescribing replaces the paperwork and allows the doctor to communicate directly to the pharmacy through database. The EHR mainly improves the safety of medication orders, but more modern system allows electronic ordering of tests, procedures, and consultations as well. EHR can be integrated with a clinical decision support system (CDSS), which can act as error prevention tool through guiding the prescriber on the preferred drug doses, route, and frequency of administration [7–10].

Cheaper health care

With increase in the application of information technology, the cost of health-care services will come down. HIT can streamline claims processing, cut costs by a large margin, and significantly improve turnaround times. The technology can automate data collection and communication process while proving the events that lead to a claim, resulting in lower health-care costs as well as a better customer experience. Due to automation of the entire HIT process data collection and communication process will be shorter with cost reduction.

Focus on preventing diseases

It is well understood that keeping a person healthy is much easier and less expensive than treating an ill person. For example, preventing diseases such as diabetes or even something like a heart attack is less expensive than a patient under treatment, without any guarantee. Since it is significantly cheaper to prevent diseases rather than curing it, governments across the world are spending a large amount of money earmarked for health care toward health information technology specifically. This would be immensely helpful in bringing down the costs of health care besides increasing efficiencies over a period of time. Developing countries especially can benefit a lot from health-care information technology as it can help in preventing outbreaks of major diseases that are plaguing the country.

Remote patient monitoring (RPM)

Remote patent monitoring (RPM) systems are gaining popularity in both urban and rural areas. This method is helpful in communicating patent data from anywhere to health-care professionals at the facility. The data enable doctors to predict and prevent any event that otherwise may cause serious issues. The RPM can help reduce the overall cost for health services and provide for better quality of health care. It also reduces the cost of chronic disease and assists in postdischarge care as well. It can also help monitor a patient's glucose level or blood pressure from anywhere across the world. Thus, it results in improved patient care.

Master patient index (MPI)

Nowadays, hospitals in urban areas also have system facilities by which patient records from more than one database can be gathered. The data can be stored for future reference by any institute sharing the database. The MPI creates an index of all medical records for a specific patient, which is easily accessible by all departments. The main purpose of MPI is to reduce duplication of patient's records and also to avoid inaccuracy of information that can result in wrong treatment.

Patient portal

A patient portal is a website for personal health care. The online tool helps to keep track of an individual health-care provider visit, test results, billing, prescriptions, etc. One can email to provider about the problems related to health care and solutions instantly. Many providers now offer patient portal. For access one can set up an account with free service. In this connection a password is used so that all of the patient information is private and secure. Besides this, patient portal provides a platform where patients can access their health-related data using any device. It includes all the information stored in an HER, such as patient's medical history, treatment, and other medications previously taken. In fact, some of the patient portals even facilitate patient to have a conversation with health-care professionals. So, instead of having to wait in line for hours to schedule an appointment, a patient can now simply log in, check their doctor's availability, and, also, have a look at their reports. But many hospitals and private practitioners are skeptical about adopting this new biotechnology network scheme because of high budget involvement and lack of well-trained person to handle HIT.

Clinical decision support

Clinical decision support provides the health-care professionals with updated information and patient-specific information. Reports based on this method save the time of doctors and assist in proper and quick diagnosis. It is also helpful in notification, alerts, and reminders to care providers and patients, clinical guidelines, condition-specific order sets, patient-specific clinical guidelines, documentation templates, investigations, and diagnostic support [11,12].

Electronic sign-out and hand-off tools

This process is referred to the passing patient-specific information from one caregiver to another by ensuring patient care continuity and safety [13]. Electronic sign-out applications are the method used as a standalone method or integrated with electronic medical record to ensure a systematic transfer of patient information during health-care provider handoffs [14–16].

Electronic incident reporting

This is a web-based technology that permits health-care providers who are involved in safety events to voluntarily report such incidents that can be helpful in integrating the events with electronic health record (HER) to enable abstraction of data and automated detection of adverse events through trigger tools. Electronic incident is helpful in standardizing incident action flow, and rapid identification of serious incidents and trigger events, while automating data entry and analysis [17,18].

Telemedicine

Telemedicine refers to the application of telecommunication technologies to facilitate patient to provider or provider to provider communication [19]. More detail is given in section.

Clinical IT advancements

Wearable technologies (type of electronic devices that can be worn as accessories, embedded in clothing, implanted in the user's body to monitor a user's health) and IoT devices (Internet of Things = pieces of hardware such as sensors, actuators, gadgets) in clinical care are progressively advancing in clinical diagnosis. Some of the most widely used wearable devices used for evaluating and monitoring blood pressure include cuffless blood pressure sensors, wireless smart phone-enabled upper arm blood pressure monitors, mobile applications, and remote monitoring techniques. Robotic surgery developments are expected to continue across the health-care system, in areas including spine, cardiology, and oncology.

Digital front door and digitization of the consumer experience

Digital front door could help the health-care provider to keep in touch with patients in COVID-19 pandemic situation, although instruction in maintaining social distancing is strictly advised.

Predictive analysis

Predictive information extracts the patient-specific information from the electronic health record (EHR) on a timely basis for a long-range forecast or an immediate patient condition in an emergency department. Currently, health-care units prefer to use predictive information technology that is used to detect early signs of patient deterioration in the ICU and general ward, identify at-risk patients in their homes to prevent hospital readmissions, and prevent avoidable downtime of medical equipment. Recently, Mount Sinai Health System in New York City developed machine learning-powered models to identify high risk and likelihood of mortality among COVID-19 patients for more efficient patient management.

The CIO and IT team evolution

A health-care CIO (health-care chief information officer) is an executive at a health-care organization that oversees the operation of information technology department and consults with other C-level personnel on technology-related needs and purchasing decisions. In pandemic situation such as COVID-19, health system need the urgency of virtual care, remote work, and more coordinated communication and data managing and operation. In this connection CIO's expertise is in high demand.

Cybersecurity

Cybersecurity plays a significant role in operating information technology in health-care unit. A health-care organization is well organized into groups with specialized hospital information system such as HER system, e-prescribing systems, practice management supporting systems, clinical decision support system, radiology information system, and computerized physician order entry system. In addition, wide varieties of sophisticated instruments are to be protected.

Email is a primary means for communication within health organization. Email carries all types of secrete and confidential information related to health system management. So, it is utmost important to ensure email security as a part of cybersecurity in health care.

Phishing is a type of social engineering attack often used to steal user data, including login credentials and credit card numbers. In this process an attacker sends a fraudulent message to capture the information and secrete data from a computer system or to deploy malicious software on the victim's infrastructure such as ransomware. Cyber criminals are taking advantage of the COVID-19 pandemic by using widespread awareness of the subject to trick users into revealing their personal information or clicking on malicious links or attachments, unwittingly downloading malware to victimized computer owners.

Artificial intelligence

Artificial intelligence is an advanced information-based technology to analyze a wide range of health-related data for health system management. Several types of software are available to help health-care provider have a thorough knowledge on a health-care system operation condition. The key categories of applications involve diagnosis and treatment recommendations, patient engagement and adherence, and administrative activities.

In COVID-19 pandemic artificial intelligence (Al) became crucial in developing predictive models for COVID-19 cases spreading across the world. Many premier world institutes and universities such as New York City-based Mount Sinai, Johns Hopkins; Rochester, Minn.-based Mayo Clinic; and University of California-Irvine developed tools and models to track the virus and estimate the risk of COVID-19 patients to track the virus and estimate the risk of COVID-19 patients developing severe symptoms.

4.3 Types of health-care information technology

Following are different types of health-care information technology (Fig. 4.4).

4.3.1 Medical practice management system

A medical practice management system refers to a type of health-care software that manages the day-to-day operations of a clinic, such as appointment scheduling, billing, and other administrative tasks. It is an integral part of the health-care system. Basically, medical practice management software can take care of insurance claim instead of patient' medical data. It automatically processes claims, handles billing and payment, and also generates reports. The software can be applicable to all health-care providers from single doctor practitioner to a large hospital. This software can also be helpful to physicians and office staff to quickly book and confirm appointments and manage schedules across multiple providers, location, and days of the week.

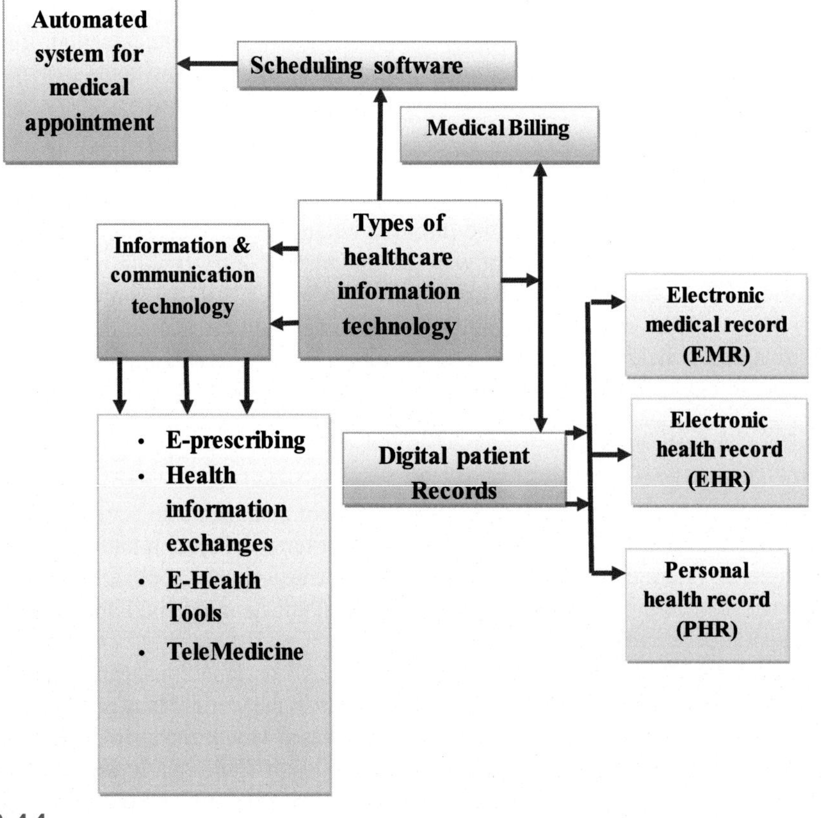

FIG. 4.4

Types of health-care information technology.

4.3.2 Appointment scheduling

Medical practice management systems allow physicians and administrative staff to quickly book and confirm appointments and manage schedules across multiple providers, location, and days of the week. Even patients can book their own appointment online. Generally, a medical practitioner faces hurdle in scheduling to attain, simultaneously, conference call and patient visit and other miscellaneous meetings. Medical practice management systems give staff member's real-time insight into each provider's availability. The software also helps clinical staff avoid double booking.

4.3.3 Organizing patient data

By this software, the detail of about patient information including extensive clinical data, such as medical history, and medications of past visit can be well documented. It can automatically verify a patient's insurance coverage prior to the patient's visit.

4.3.4 Billing and claims processing

This software is useful in preparing billing statement and generating electronic claims based on patient encounters, on the basis of payer regulations and correct diagnosis codes. If the medical practice management system is cloud based, new billing and procedural codes and rules will automatically update, ensuring that the practice never uses outdated codes.

4.3.5 Reporting

The overall performance and running of personal clinic, hospital departmental activities, and even overall performance and efficiency of hospital management system can be updated through HIT. It can also provide update on the following on a regular basis:

- How often patients fail to show up for their appointment?
- How much time do staff members spending on tasks?
- How quickly payers reimburse claims?
- How the clinic is performing against meaningful use objectives?
- How long has a claim been in accounts receivable?

4.3.6 Urgent care applications

It is a type of health management information system that brings awareness within patients who require immediate attention. It provides updated information about health-related queries, informative health articles, and even keeps track of their medical status.

4.3.7 Medical billing software

Medical billing preparation is one of the most time-consuming and integral part of medical practices and services. Generally, health-care practitioners are extremely busy with shortage of time for administrative and financial settlement jobs. Billing software makes it easy by automatically generating medical bills and handling the entire workflow. Besides patient billing, the software also takes care of insurance claims and verification, payment tracking, and processing.

4.4 Health services in rural communities

Rural population refers to people living in rural areas as defined by national statistical office. It is calculated as the differences between total population and urban population. Half of the global population lives in rural areas in developing countries. Due to the lack of communication infrastructure and unavailability of resources it has been difficult to extend information and communication technologies (ICTs) in rural and remote areas. However, several initiatives have been taken to develop low-cost computer, wireless communication infrastructure, and open-source software to promote ICT in rural sector for developing sustainable pattern of health services. For the development of sustainable ICT in rural and remote locality, a state should put effort in reshaping education, socioeconomic condition and development of rural infrastructure [20]. Plenty of feedbacks have been available in the literature, since the last two decades [21–24].

Implementation of frame work of ICT would immensely help in rural sector in improving health-care services (Fig. 4.5). But to improve e-Healthcare, there are also challenges in implementing ICTs [25–28]. Rural areas and urban communities are different from health perspective (Fig. 4.6A,B).

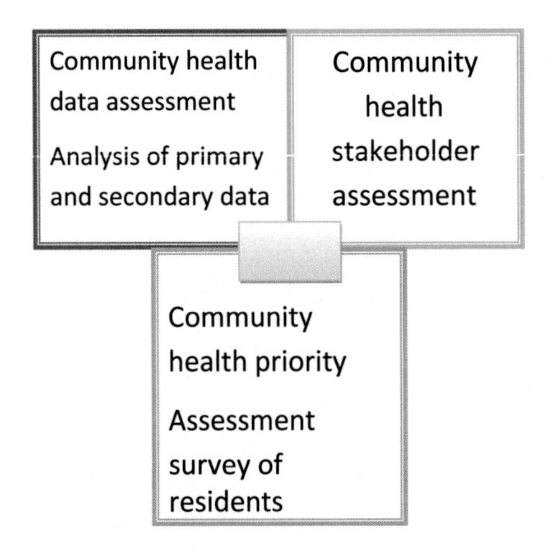

FIG. 4.5

Challenges for initiating rural health-care development.

Slum area near railway track New Delhi, India **Slum area near sewage canal,Mumbai, India**

FIG. 4.6

Habitable life of poor people residing near unauthorised colony nearby railway track, New Delhi (A) or sewage canal, Mumbai (B), India.

Both the rural and the urban locality can be projected differentially from their health-promotion behaviors, health maintenance, and illness treatment. The rural and remote areas in developing countries mostly lack health-care workers and availability of health-care infrastructure. The poverty and long distances between health-care providers and rural areas are mainly responsible for the provision inadequate health services.

The public health-care center (PHC) and subcenter in rural areas are mostly not well equipped with infrastructure and staff to provide quality health care to the rural poor. The modernized health-care services have not yet been percolated to the rural areas, and this is a matter of great concern. So, it is high time to explore the possibility of bringing equity of access to health professionals and institutions between rural and urban areas to bring sustainability in rural health services (Fig. 4.7).

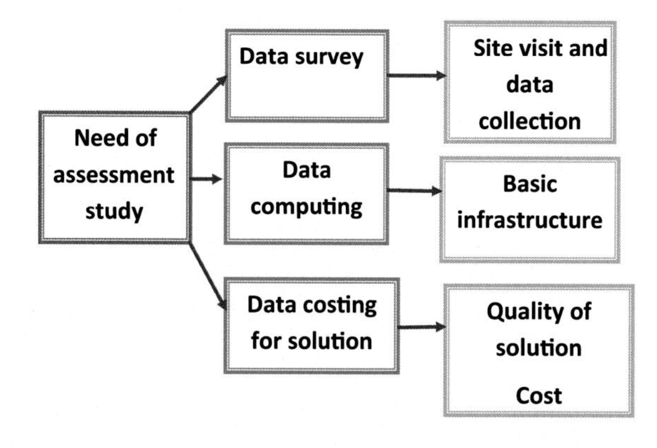

FIG. 4.7

Schematic diagram of bringing sustainability in rural health-care development.

Literature review suggests that rural hospitals are in the worst financial conditions due to several factors. They are smaller, located in remote areas, and provide less specialized services due to their problems with employing well-qualified staff. The concept of rural sustainability is supported by good financial conditions of rural hospital, which helps to provide better access to medical services for inhabitants of rural areas (Fig. 4.8).

4.4.1 Life expectancy

In rural area, it has been noted that in many parts of the world life expectancy rates are higher in urban areas as compared to that in rural and remote localities [29]. It could have been due to the wide gap existing in education system and economic condition between urban area and rural locality [30].

For example, in Canada, the life expectancy in men ranged from 74 years in the most remote areas to 76.8 years in its urban center. For women, life expectancy was also lowest in rural areas, with an average of 81.3 years. Those living in rural areas adjacent to urban centers also experience highest life expectancies (with men at 77.4 years and women at 81.5 years). Australian life expectancies ranged from 78 years in major cities to 72 years in remote locations [31]. In China, the life expectancy of female is 73.59 years in urban area and 72.46 years in rural areas. Male life expectancy varies from 69.73 years in urban areas to 58.99 years in rural areas [32].

But, reverse trend to above facts is noted in United Kingdom where life expectancy in rural areas exceeds that of urban areas.

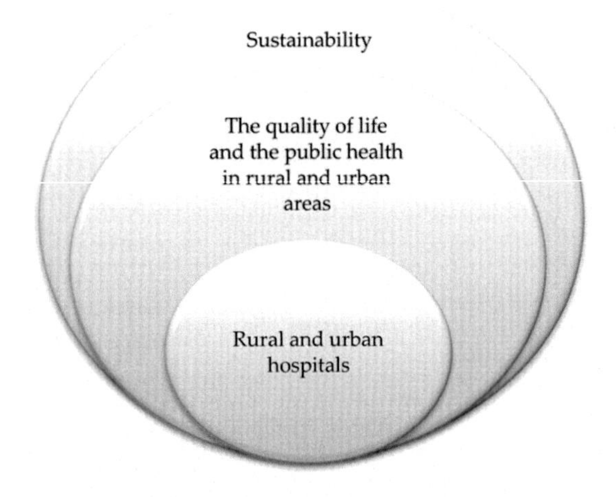

FIG. 4.8

Rural health-care services and sustainability.

4.4.2 Health determinants in rural areas

People in rural communities have less provision to health care than their urban counterparts. Health-care workers face inconvenience in discharging health-related services in rural communities due to the lack of internet facilities, adequate IT, and mobile preventive care and treatment programs. Various health determinants have already been discussed in the earlier section of this chapter. Health funding framework is not to the expectation in both developed and developing countries. In 1993, in China only 10% of the rural population had medical insurance. But, at present about 95% of the population has at least basic health insurance coverage. The United States primarily has a third-party payer system of health care, which means that a health insurance plan (the third party) reimburses doctors for the bulk of the cost of health-care services provided to patients. The nation used a mixed system of public and private insurance.

Most of the underdeveloped countries such as Africa, have inadequate health-care recruitment practice which resulted in less number of health-care professionals over the desire number needed for satisfactory health-care practices. In sub-Saharan Africa, urban and more prosperous areas have disproportionately more of the countries' skilled health-care workers. For example, urban Zambia has 20 times more doctors and over five times more nurses and midwives than the rural areas. In Malawi, 87% of its population live in rural areas, but 96.6% of doctors are found in urban health facilities. In South Africa alone, half of their population lives in rural areas, but only 12% of doctors actually practice there [33–35].

4.4.3 Rural health care and WHO

Since 1992, World Organization of National College and Academy (WONCA), a World Organization of Family Doctors, has developed a specific focus on rural health through the WONCA working party on rural practice. This working party has drawn national and international attention to major rural health issues through World Rural Health Conferences and WONCA rural policies. WONCA is a global not-for-profit professional organization representing family physicians and general practitioners from all regions of the world. It has official relation with World Health Organization (WHO). In April 2002, WHO and WONCA held major WHO-WONCA Invitational Conference on Rural heath. This conference addressed the immense challenges for improving the health of people of rural and remote areas of the world and initiated a specific action plan: The Global Initiative on Rural Health.

In spite of wide of gap between developing and developed countries, access is a critical issue in global rural health problem. Even in the countries where the majority of the population lives in rural areas, the resources are concentrated in the cities. Invariably, all countries have difficulties with transport and communication, and shortage of doctors and other health professionals in rural and remote areas.

4.5 Rehabilitations during COVID-19 and telehealth

The past pandemic scenario of COVID-19 and present slow retardation process have resulted due to accurate diagnosis and effective isolation and intense care for all cases of COVID-19 including cases with mild or moderate disease. During advanced stage of COVID-19, the primary attention was given to rapid identification, testing, and treatment of patients contaminated with COVID-19. In addition, temporary health-care shelters were developed to isolate the needy COVID-19 patients who have poor housing facility. COVID-19 resulted in significant challenges for rehabilitation services around the world. Rehabilitation services in COVID-19 have been consistently disrupted. At the same time, demand for the rehabilitation of patients in temporary shelter has increased both for patients who are critically unwell with the disease and for those who continue to experience the long-term consequences of their illness. WHO, together with partners, has issued guidance, and provided resources to tackle the pandemic situation caused due to COVID-19. This includes integrating rehabilitation into key clinical guidelines and developing stand-alone rehabilitation resources. Patents ventilated for longer time due to critical illness usually require intense rehabilitation input as more than 60% are unable to walk. In case a patient experiences a stroke or cardiac complications following COVID-19, he/she will require rehabilitation for an extended period, and in some cases, they will require lifelong support [36].

It is extremely necessary that COVID-19 patients are provided health-care workforce [37], but due to mandatory provision of social distancing, health-care providers need to explore innovative approaches for rehabilitation to ensure that both COVID-19 and non-COVID-19 patients receive adequate care and support.

4.5.1 Telehealth

An inadequate rehabilitation service has significant impacts on patients, families, and health workers. Rehabilitation provider can adapt telehealth or virtual rehabilitation [38]. Telehealth refers to the delivery of health-care services via electronic communication (phone, internet, video calls) [39]. Telehealth can provide preventive; promotive, and curative aspects of health and many different health-care professions are involved in its delivery (Fig. 4.9) [40].

Often, people referred telehealth as remote clinical services such as diagnosis and monitoring. Telehealth is especially helpful to the people living in rural community where inadequate transport service persists with lack of mobility, poor funding from government, nonavailability of primary health center, and lack of health workforce (Fig. 4.10).

Under such condition, telehealth may bridge the gap by providing online information and health data presentations between practitioners and clinical discussion through video conference. Besides this, telehealth may be helpful in robotic surgery occurring through remote access; physical therapy done via monitoring instruments, live feed, and application combinations; tests being forwarded between facilities for

FIG. 4.9

Telehealth service being used by a doctor for prescription and advice.

TELEHEALTH SERVICES MODALITIES

Interactions	Devices	Modalities	Patient Location
• Patient to provider • Provider to provider	• Smartphone • Computer/tablet • Monitoring device	• Videoconference • Remote patient monitoring • Phone* • Secure messaging*	• Home (or location of choice) • Clinic/Office • Hospital

FIG. 4.10

How telemedicine facilitate a broad range of interaction using different devices and modalities.

interpretation by a higher specialist; home monitoring through continuous communication of patient's data; client to practitioner online conference; or even videophone interpretation during a consult [41].

The benefits of virtual rehabilitation are as follows:

- The user feels elevated by using innovative technology and become more friendly with doctor with less likelihood that the patient will stop treatment.
- Direct contact between patients and doctor or health-care workforce develop more trust of patients on doctors, and helpful in bringing peace of mind.

There are certain challenges associated with telehealth which may limit its uptake prior to COVID-19 [1]. Both the health-care provider and patient should possess the knowledge of handling telehealth technology and able to access support for any trouble shooting. It is also essential that providers explore if they are licensed/insured to practice telehealth and if they will be reimbursed for their work [42,43].

4.5.2 Telehealth vs telemedicine

Sometimes, telemedicine is used as synonym of telehealth. But, the Health Resources and Services Administration (HRSA) have differentiated telehealth from telemedicine. As defined by HRSA telemedicine is only means for remote clinical services, such as diagnosis and monitoring, while telehealth includes preventive, primitive, and curative care delivery [40,44].

The US Department of Health and Human Services referred the term telehealth as "nonclinical services, such as provider training, administrative meetings, and continuing medical education," and the telemedicine means "remote clinical services."

The World Health Organization uses telemedicine to describe all aspects of health care including preventive care [45]. Telemedicine and telehealth are used interchangeably by the American Telemedicine Association [46].

Store and forward

The process of acquiring medical data related to medical images, biosignals, and subsequently transferring to medical practitioners at a convenient time for assessment off-line is most critical and important practice for health-care providers. It is one of the primary duties of health-care providers. It does not require the presence of both parties at the same time. In this connection, telemedicine is immensely helpful to dermatologist, radiologist, and pathologist who are well versed with telemedicine. The image and data transition with specialist is mostly made with electronic medical report and audio/video information in lieu of a physical examination.

Remote monitoring

Remote monitoring, otherwise known as self-monitoring or self-testing, is useful to medical professional to monitor a patient remotely using various technological devices. This method is basically used for managing chronic diseases or specific conditions, such as heart disease, diabetes mellitus, or asthma. These practices are comparatively convenient as compared to traditional in-person patient encounters, supply greater satisfaction to patients, and may be cost effective.

Real-time interactive

Currently, real-time interactions between patient and health-care provider through electronic consultation is more convenient and time saving. Videoconferencing has been used in a wide range of clinical disciplines and settings for various purposes including management, diagnosis, counseling, and monitoring of patients.

Videotelephony

The audio-video signals of patient's health issues can be easily and safely transmitted to health-care specialists from different locations, for communication between people in real time. Through videotelephony technology both still images and real-time video of patients can be communicated to specialists instantly.

Currently, videotelephony is useful to deaf and speech impaired patients who can use them with sign language or those who are located in distant places and are in need of telemedical or tele-education services.

4.5.3 Telehealth for patients in the hospital

In order to limit the spread of the virus, some hospitals have adopted telehealth technology within the hospital to do virtual therapy sessions. So, the COVID-19 patents should be provided an iPad, stand and speaker, and well trained to communicate their respective health issues with health-care provider easily. The therapist could observe the patient doing tasks such as sitting up, reaching, and putting on socks and give them advice and technique to decrease fatigue. Real-time feedback can also be provided when a computer program monitors the patient during movement through motion detection and gives feedback as the patient moves. For example, a Hasharon hospital in Israel is devoted solely to corona patients. The physiotherapy department has taken the responsibility for providing services specifically framed for patients with COVID-19. This program is being coordinated with well-trained health workforce to meet the challenge of patient's specific needs with the help of telehealth practice. In addition, they prepare exercise sheets and videos in multiple languages [47].

Telehealth is proven to be the most important practice in hospital to minimize direct exposure of the health-care provider to COVID-19 patients. It is the suggestion of multidisciplinary team that the physiotherapists should avoid long-term exposure to COVID-19 patents

- with mild to moderate respiratory symptoms
- who can independently clear secretion

4.5.4 Telehealth services

Emergency service

To meet the urgency of emergency service in corona diseases 2019 pandemic, separate emergency department (ED) physicians, staff, and health-care system with a myriad of challenges were developed. EDs have been tasked with addressing the acute care needs of patients. Many EDs have leveraged technologies such as telehealth to innovate and treat patient safely. A variety of tools such as telephone, smartphone, and mobile wireless devices are used in telehealth.

Medication-assisted treatment through telemedicine

Medication-assisted treatment (MAT) is referred to the treatment of opioid use disorder (OUD) with medications, often in combination with behavioral therapy. The Drug Enforcement Administration has approved buprenorphine (trade name Suboxone) via telemedicine without the need for an initial in-person examination. On March 31, 2020 QuickMD became the first national Tele-MAT service in the United States to provide medication-assisted treatment with Suboxone online without the need for an in-person visit.

Telenutrition

It is practice of online counseling for taking nutrients or nutrient supplements to maintain a healthy lifestyle. Patients can upload their vital statistics, diet logs, food pictures, etc., on telenutrition portal, which are then used by nutritionists or dietitians to analyze their current health condition. Through telenutrition portal client/patients can get remote consultation for themselves and their family from the best nutritionists or dieticians available across the globe. Even during COVID-19 lockdown period clients/patients can easily avail the facility of telehealth for health management from home, without having to visit the doctor.

Telenursing

Under telehealth program telenursing service can be provided wherever a large physical distance exists between patient and nurse, or between any numbers of nurses. For homecare services telenursing can be most effectively useful for the patients who are immobilized, live in remote or difficult to reach places, or who have chronic ailments, such as chronic obstructive pulmonary disease, diabetes, congestive heart disease, and Alzheimer's disease. Telenursing can be a communicating bridge between nurse and at home patients through video conference, internet or videophone. Telenursing has been gaining appreciation and acceptability in many countries due to preoccupation in reducing the costs of health care, an increase in number of aging and chronically ill population, and the increase in coverage of health care to distant, rural, small, or sparsely populated regions. Telenursing may be useful in increasing shortage of nurses, to reduce distances and save travel time, and to keep patients out of hospital.

Telenursing can play a significant role in preventing, diagnosing, treating, and controlling severity of disease during COVID-19 outbreak. Telenursing can also be, especially, helpful when people are in quarantine, enabling patients in real time through contact with nurses for advice on their health problem.

Telepharmacy

Telepharmacy is the delivery of pharmaceutical care via telecommunication to patients in remote rural communities (Fig. 4.11).

FIG. 4.11

A pharmacist using internet facility to provide prescribed medicine and advice.

It is an instance of the wider use of telemedicine, as implemented in the field of pharmacy. Telepharmacy services include drug therapy monitoring, patient counseling, prior authorization and refill authorization for prescription drugs, and monitoring of formulary compliance with the aid of teleconferencing or videoconferencing. People can avail telepharmacy services at retail pharmacy site, or through hospitals, nursing homes, or other medical facilities.

Telerehabilitation

Telerehabilitation is the practice of delivery of rehabilitation services over telecommunication network and internet. Through telerehabilitation patients can interact remotely with health-care providers and can be used both to assess patients and to deliver therapy.

Telerehabilitation refers specifically to clinical rehabilitation services with the focus on evaluation, diagnosis, and treatment. Telerehabilitation can be provided in many alternate forms such as real-time visit with audio, video, or both; asynchronous e-visits; virtual check-ins; remote evaluations of recorded videos or images; and telephone assessment and management services.

Through telerehabilitation a variety of fields of medicine such as physical therapy, occupational therapy, audiology, and psychology can be handled during COVID-19 pandemic situation. Therapy sessions can be individual or community-based. Types of therapy available include motor training exercises, speech therapy, virtual reality, robotic therapy, and group exercise. Commonly used modalities include webcam, videoconferencing, phone lines, videophones, and webpages containing rich internet applications.

COVID-19 has presented challenge for finding out some alternate service to resolve health-care risks due to the uncontrolled severity of the disease. In response, telerehabilitation has emerged as an alternative care model to minimize and control the severity of the pandemic.

Currently, intensive research is in progress on new and emerging rehabilitation modalities as well as comparisons between telerehabilitation and in-person therapy in terms of patient functional outcomes, cost, patient satisfaction, and compliance.

Telerehabilitation can deliver therapy to people who cannot travel to a clinic because the patient has a disability or because of long travel time. Most telerehabilitation is highly visual. The most commonly used mediums are webcams, videoconferencing, phone lines, videophones, and webpages containing rich web applications. It is most commonly used for neuropsychological rehabilitation; fitting of rehabilitation equipment such as wheelchairs, braces, or artificial limbs; and in speech-language pathology. In 2001, for the first time cognitive impairment were introduced for neuropsychological rehabilitation.

Teledentistry

The use of information technology for dental care, consultation, education, and awareness on dental hygiene is known as teledentistry. As the COVID-19 spreads by droplet, fomite, and construct transmission, face-to-face interaction of health-care professional with patient carries a risk of its transmission. Due to close contact of dentist with patient, it has been advised not to go for dental clinical service, unless the case is supposed to be in critical condition.

Teleaudiology

Teleaudiology is the utilization of telemedicine to provide audiological services to the patient. There are two types of teleaudiology tests: (i) the patient is tested and data related to testing is then store-and-forward to specialist for advice and treatment and (ii) the other alternate test is real-time test through tele-conference by remote device.

Teleneurology

This test is mostly conducted through mobile technology to provide neurological care remotely, including care for stroke, movement disorders like Parkinson's disease, seizure disorder (e.g., epilepsy), etc. Two main techniques used in teleneurology are: (i) videoconferencing, which enables communication between a doctor and a patient who are in different places, (ii) through email, where the consultation is carried out without the presence of patient, at the time of convenience of the doctor involved.

Teleneuropsychology

Teleneuropsychology is the use of telehealth/videoconference technology for the remote administration of neuropsychological tests. The cognitive status of a patient with brain disorders can be investigated through teleneuropsychological tests.

Telepsychiatry

Telepsychiatry, also known as telemental health or e-mental health, is broadly defined as the use of ICT to provide or support psychiatric services across distances. It is a type of telemedicine therapy utilizing videoconferencing for the patients residing in remote areas or peripheral location of developing countries. Telepsychiatry can provide easy way to clients to access psychiatric services if they are unable to travel. Using advanced technology, mental health-care providers can extend their specialized services to clients in remote rural areas and peripheral location of developing countries. Such technology has been used for psychiatric consultation, assessment, and diagnosis, medication management and management in individual and group psychotherapy.

4.5.5 The practice of telemedicine

The practice of delivering telemedicine broadly broken down into three types: (i) store-and-forward, (ii) remote patient monitoring, and (iii) real-time encounters.

Store-and-forward telemedicine

Store-and-forward telemedicine is also known as "asynchronous telemedicine." In this practice, health-care providers collect and share patient medical information such as laboratory reports, images, video, and other record with a physician, radiologist, or specialist at another location. It is not similar to email, but it is done using a solution that has built-in, sophisticated security feature to ensure patient confidentiality. Patients, primary health-care provider, and specialists should collaborate with mutual understanding to increase the efficiency of Store-and-forward telemedicine system. Store-and-forward telemedicine is acceptable to all for diagnoses and treatment with certain specialties including dermatology, ophthalmology, and radiology.

Remote patient monitoring or "telemonitoring" is to be in touch with patient's vital sign and activities, especially with high-risk patients, like those with heart conditions and people gets relieve, immediately from hospital. Remote patient monitoring is also extremely useful for the treatment of a number of chronic conditions. It can be used by diabetic patients, for example, to track their glucose levels and communicate the data to their respective physicians.

Real-time telemedicine

In real-time telemedicine both patients and health-care providers can have discussion about the status of telemedicine delivery through videoconferencing. While the other types of telemedicine are used to enhance traditional in-person visit, real-time telemedicine can be used in lieu of a trip to the doctor's office under certain situation of emergency. It is popular for primary health care, urgent care, follow-up visits, and for the management of medications and chronic illness.

It is never advisable to use any consumer video communication that is used to connect with friends and coworkers, such as Facetime and Skype, in telemedicine practice. Telehealth encounters should be conducted using technology that has been designed to protect patient privacy and strictly meet the patient full protections as stated in the Health Insurance Portability and Accountability Act (HIPAA).

The first reference to telemedicine in the medical literature appeared in 1950 [48]. It was about the transmission radiologic images by telephone between West Chester and Philadelphia, Pennsylvania, a distance of 24 miles [49]. Building on this early work, Canadian radiologist at Montreal's Jean-Talon Hospital created a teleradiology system in the 1950s [50–52].

With the advancement of communication technology, in 1950, a Canadian radiologist reported diagnostic consultations based on fluoroscopy images transmitted by coaxial cable. Subsequently, health workers from around the globe started reporting on telehealth [52–57].

The COVID-19 pandemic has challenged the health risk service, and brought about many changes and projected telehealth into the mainstream of health-care delivery. Audio and video conference health care visits have become commonplace and have impacted geographic barriers and access to care issues with the potential for care coordination in the fragmented health-care delivery system.

4.5.6 Advancements in telehealth technology

In the 1970s, the telemedicine was in its infancy with the limited use of old traditional technology and expensive videoconferencing technology to connect local providers with remote specialists. Although, this type of real-time communication is still used, advancements in communication technology have made its use much simpler and more affordable. The modern videoconferencing system gives the feeling of a discussion sitting before doctor, in a common room. These types of systems can also be utilized to provide patients with quality counseling or other mental health care that may be unavailable where they live. With the advancement of wireless communication devices major improvement in telehealth system has occurred [58]. This allows patients to self-monitor their health conditions and to not rely much on health-care professionals. These modified and modernized technologies make the patients friendlier with health-care provider and develop confidence on treatment process [59,60].

During this COVID-19 pandemic, modernized telehealth has emerged as effective and sustainable solution for precaution, prevention, and treatment to minimize the spread of the disease. The latest telehealth technology has been playing a significant role in bridging between health care providers and health-care system by giving solutions to the challenges of health risk issues in COVID-19.

4.5.7 Licensing

US licensing and regulatory

To deliver telemedicine care, a physician has to obtain legal permission under restrictive licensure laws that vary from state to state. In case a practitioner serves several states, then he/she has to obtain license from each state, which could be an expensive and time-consuming task.

In case a practitioner never practices medicine face to face with a patient in another state, he/she still must meet a variety of other individual state requirements, including paying substantial licensure fees, passing additional oral and written examinations, and traveling for interviews. In this connection, the United States has already passed the Ryan Haight Act, which required face to face or valid telemedicine consultations prior to receiving a prescription [61].

The state medical licensing boards have sometimes opposed telemedicine, but in 2015 the state legislature legalized electronic consultations.

4.5.8 Major implication and impact

A variety of information technology has been merged with telehealth care in order to make the system more potential and uniform level of care. As telehealth enhance mainstream health care, it challenges notions of traditional health-care delivery.

Health promotion

The last decade has witnessed the pattern of health promotion, both in rural and in urbanized areas. Remaining at home the patients from various localities can safely avail quality telehealth facilities. Health promotion has been gaining popularity where there are very poor physical resources available. Mobile health (mHealth) can be easily applicable in remote areas in underdeveloped countries.

In developing countries health promotion through telehealth is gaining acceptability. In 2014, Australia Breastfeeding Association helped new mothers learn how to breastfeed. World Health Organization strongly recommended breastfeeding, which is highly beneficial to both the mother and the baby (Fig. 4.12). Breastfeeding, also called nursing, is the process of feeding human breast milk to a child, either directly from the breast or by expressing (pumping out) the milk from the breast and bottle feeding it to the infant.

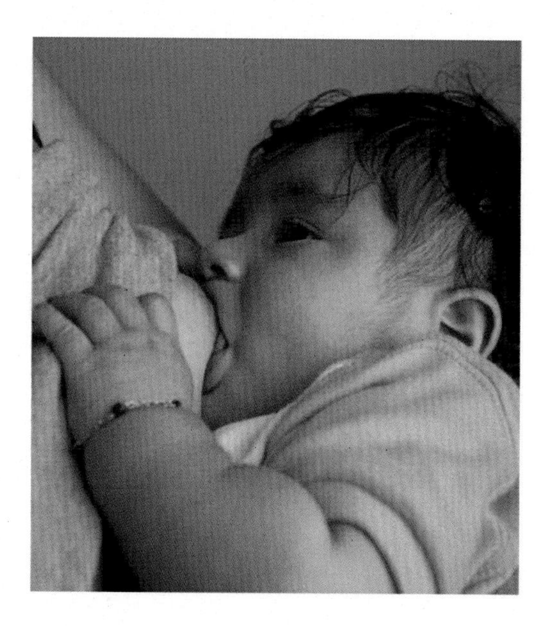

FIG. 4.12

Practice of breast feeding to children.

Health-care quality

The desired quality health outcome primarily depends on many aspects including telehealth care serve, especially during emergency of outbreak of pandemic such as COVID-19. As countries commit to achieve "Health for ALL," it is imperative to carefully consider the quality of care and health services. Quality health care should be effective, safe, and people-centered. It should be delivered timely, equitable way through integrated service system.

4.5.9 Economic evaluation

It is necessary to assess economic evaluation before implementing telehealth care system in a specific geographic location. The economic analysis requires the assessment of costs and benefits associated with telehealth keeping in view the infrastructure availability and adequate budgetary provision of the state government.

It is primarily important to understand whether overall cost of telehealthcare infrastructure development will give quality outcome on health promotion. It is necessary to explore economic evaluation studies of telemedicine in terms of their descriptive analysis, methodological quality, and over all outcome of the project. It is also necessary to follow well-standardized frame work for economic evaluation, with a view for providing tool for policymakers and stakeholders. The overall economical viability relies of telemedicine depends on the funding model within the country, the consumers' willingness to pay, and the expected remuneration by the clinicians or commercial entities providing the services.

4.5.10 Limitations and restrictions

Telehealth does not work as universal filler to many branches of medicine. There are certain risks and barriers that bar the full merging of telehealth into best practice. Although it is predicted that telehealth will replace many consultations and other health interactions, it cannot fully replace a physical examination such as diagnostics, rehabilitation, or mental health.

The benefits of telehealth is limited to certain extend due to legislation and practice. It is obvious that telehealth serves the clients timely. It is obvious that telehealth serves the clients timely, wherever they may be, although this is a benefit, it also poses threat to privacy, safety, medical licensing, and reimbursement. In case the clinician and patient are in different geographic locations, it is difficult to determine which laws apply to the context. For example, in United States, once health care crosses borders of different state bodies are decision factor for allowing the health-care providers to exercise their service for promoting health system. For example, in United States, once health-care provider crosses borders of different state. The decision factor for allowing the health-care providers to exercise their service for promoting health system depends on the policy of the respective state. Medicare will only reimburse live video (synchronous)-type services, not store-and-forward, m-health, or remote patient monitoring (it does not involve live video).

It is necessary that more specific and widely reaching laws, legislations, and regulations evolve with the modernization of telemedicine health-care facilities.

Appropriate clinician-patient relationship can be established online to make prescribing safe. It may be necessary that before any telemedicine practice transition the clients and license holder health-care providers should meet personally, at least once before online prescribing can occur, or that at least a live-video conference occurs.

4.5.11 WHO consent on telemedicine

Information and communication technologies (ICTs) have potential to meet various health-care challenges in providing accessible, cost effective, and high-quality health-care services under emergency situation such COVID-19 pandemic. Telemedicine uses ICTs to overcome geographic barriers, and increase access to health-care services. This is especially the case for rural areas of developing countries.

Keeping in view the great potential of telemedicine for health promotion, in 2005, the World Health Organization established Global Observatory for eHealth, and urged Member States "to consider drawing up a long-term strategies plan for development and implementing eHealth services—to develop the infrastructure for information and communication technologies for health—to promote equitable, affordable and universal access to their benefits." Countries and stakeholders were urged to direct their efforts toward creating a consistent eHealth vision in line with a country's health priorities and resources, developing an action plan to deliver the proposed vision, and creating a framework for monitoring and evaluating eHealth implementation and progress.

The main objective of Observatory is to determine the various issues related to electronic Health (eHealth), including telemedicine, at the national, regional, and the global level, and providing the WHO's Member States with reliable information and guidance on best practices, policies, and standard in eHealth. In 2005, following the formation of WHO's eHealth strategy, the Observatory conducted a global eHealth among the Member States. On the basis of this survey report, the observatory started second eHealth strategy in 2009 to understand the eHealth issues in more detail. The eHealth series is primarily meant for government ministries of health, information technology, and telecommunications, nongovernmental organization, and donors. According to WHO 2018 report, over 63% of countries have implemented digital health polices and national strategy. The digital health age is in the process of executing access to health services, providing better diagnostics and health information to the individual. ICT and digital health solution can immensely be helpful in busting SDG3 target by the end of 2030. Due to the COVID-19 pandemic, the 73rd World Health Assembly was held between May 18 and 19, 2020. The World Health Assembly is the decision-making body of WHO. One of the main objectives of this assembly was to understand the global strategy on digital health to promote health for all.

Advancement in information technology has made telemedicine more effective and acceptable. So, an ethical issue in this regard is more critical for its acceptability at social level. Currently, social acceptability of telemedicine and ethical issue is a burning topic in order to bring moral confidence in the clients, seeking for telemedicine under emergency situations such as COVID-19 pandemic [62,63]. Moreover, telemedicine is a better medicine to resolve many health issues for senior citizens and those are residing in rural community lacking health infrastructure and proper transport system.

Ethical issues in telemedicine give opportunity to patients to accept telemedicine with utmost confidence and also express their dissatisfaction with the services [64]. The initiation of ethical issues started in 1980s when telemedicine was at the climax of social acceptability. In 2006, the American Society for Bioethic and Humanities (ASBH) was officially assigned by WHO to investigate ethical issues in telemedicine [65]. Tremendous use of internet system by adults, and simultaneously accelerating rate of receiving online health-related information have highlighted the role of ethical issues in telemedicine [66]. The widely proliferation of internet, email, and smart phones has further highlighted the critical role of ethical issues in telemedicine [67]. So, it is high time to set ethical rules in order to safe guard the safety and security of telemedicine practice [68].

4.6 HIT resources specific to rural facilities

Many rural specific HIT resources are available in the world. For example, the following are few important HIT resources in the United States involved in promoting health-care services in rural areas.

The Office of National Coordinator for HIT

The Office of the National Coordinator for Health Information Technology (ONC) is a staff division of the Office of Secretary, within the US Department of Health and Human Services. ONC provides an HIT Playbook that acts as a guidebook for health-care professionals for using HIT, to understand the benefits of HIT in rural areas, barriers to HIT adoption, implementation of resources, and tools and resources for critical access hospitals (CAHs) and other rural hospitals.

The National Rural Health Resource Center (NRHRQ)

It is a nonprofit organization within the United States dedicated to sustaining and improving health care in rural communities. NRHRQ provides technical assistance, information, tools, and resources for the improvement of rural health care. It serves as a national rural health knowledge center and strives to build state and local capacity.

The Agency for Health-care Research and Quality (AHRQ)

It is one of 12 agencies within the US Department of Health and Human Services. The agency is headquartered in North Bethesda, Maryland, a suburb of Washington, DC. It serves for supporting research, demonstration projects, and evaluations; developing guidelines; and disseminating information on health-care services and delivery systems.

The National Rural Health Association (NRHA)

NRHA is a national nonprofit professional association in the United States. The association provides leadership on rural health issues, which it attempts to carry out through education, communication, and advocacy. It maintains a list of HIT resources that provide information and connections for HIT implementation, use, and support for rural areas.

4.7 How HIT improve health-care delivery?

HIT works to ensure efficient, coordinated, and secure health-care information exchange for patients who receive health-care services from multiple providers or multiple locations. HIT helps patients to avail opportunity to engage in the provision of their health care by tracking health status, analyzing provider statement related to health promotion status. In this regard, availability of broadband facilities is a limiting factor in providing telehealth care network in rural areas. For example, the Federal Communication Commission's (FCC) 2020 Broadband Deployment Report noted that in 2018, approximately 22.3% of rural residents and 27.7% of Americans in tribal areas did not have access to fixed broadband services that met the FCC's minimum speed benchmark. Rural communities have inadequate internet and are less likely to manage personal health information online or email health-care providers.

Relationship between HIT and telehealth

HIT and telehealth functional aspects are complementary to each other. Both systems are used to promote health-care system, but in different ways. Examples of HIT include electronic health records (EHRs), personal health records (PHRs), e-prescribing, online communities, and online patient and provider communication. The Health Resources and Services Administration (HRSA) Office of the Advancement Telehealth (OAT) defines telehealth as: "*using electronic and telecommunications technologies to support and promote long-distance clinical health care, patient and professional health-related education, public health, and health administration.*"

4.8 HIT funding opportunity for rural provider

Availability of funding resources for HIT promotion in rural areas is limited. However, developed countries such as United States provide many grant and loan programs focusing on issues and initiatives that interplay with HIT, such as health-care quality, access to broadband access, capacity building, and network development.

Several federal organization and agencies such as Agency for Health Research and Quality (AHRQ), Health Resources and Services Administration (HRSA), Federal Office of Rural Health Policy, and Bureau of primary Health-care offer grant opportunities that support HIT. The US Department of Agriculture (USDA) has loan and grant programs that funds HIT initiatives. Other funding opportunities supporting rural providers and HIT exist. For example, the Universal Service Administration Company (USAC) provides assistance to health-care providers for eligible expenses related to broadband connectivity at a flat discount rate of 65%.

4.9 Security and privacy for electronic information

Generally, the developed countries have better regulatory act for the security of health information. In 1996, the United States created the Health Insurance Portability and Accountability Act (HIPAA), and framed regulation to provide a minimum standard for compliance with privacy and security of health information. On the basis of HIPAA security rule, risk assessment on telemedicine use should be performed on an annual basis. The risk assessment should focus on administrative, physical, and technological risk of the organization providing telemedicine service. It is also recommended that a third party be involved in the risk assessment to act as another set of eyes to identify any risk.

References

[1] Brailer D. The decade of health information technology, framework for strategic action r health care cost control: hope vs reality. Ann Intern Med 2009;150(7):485–9.

[2] Buntin M, Burke M, Hoaglin M, Blumenthal D. The benefits of health information technology: a review of the recent literature shows predominantly positive results. Health Affairs 2011;30(3).

[3] Goldzweig C, Towfigh A, Maglione M, Shekelle P. Costs and benefits of health information technology: new trends from the literature. Health Affairs 2009;28(2).

[4] Fiks AG, Grundmeier RW, Biggs LM, Localio AR, Alessandrini EA. Impact of clinical alerts within an electronic health record on routine childhood immunization in an urban pediatrics population. Pediatrics 2007 Oct;120(4):707–14.

[5] Smith DH, Perrin N, Feldstein A, Yang X, Kuang D, Simon SR, Sittig DF, Platt R, Soumerai SB. The impact of prescribing safety alerts for elderly persons in an electronic medical record: an interrupted time series evaluation. Arch Intern Med 2006;166(10):1098–104.

[6] Computerized Provider Order Entry [Internet] Agency for Healthcare Quality & Research. 2017. https://psnet.ahrq.gov/primers/primer/6/.

[7] Nuckols TK, Smith-Spangler C, Morton SC, Asch SM, Patel VM, Anderson LJ, et al. The effectiveness of computerized order entry at reducing preventable adverse drug events and medication errors in hospital settings: a systematic review and meta-analysis. Syst Rev 2014;3:56.

[8] Devine EB, Hansen RN, Wilson-Norton JL, Lawless NM, Fisk AW, Blough DK, et al. The impact of computerized provider order entry on medication errors in a multi-specialty group practice. J Am Med Inform Assoc 2010;17:78–84.

[9] Kaushal R, Kern LM, Barron Y, Quaresimo J, Abramson EL. Electronic prescribing improves medication safety in community-based office practices. J Gen Intern Med United States 2010;25:530–6.

[10] Strom BL, Schinnar R, Aberra F, Bilker W, Hennessy S, Leonard CE, et al. Unintended effects of a computerized physician order entry nearly hard-stop alert to prevent a drug interaction: a randomized controlled trial. Arch Intern Med 2010;170:1578–83.

[11] Clinical Decision Support (CDS) [Internet] *Office of the National Coordinator for Health Information Technology.* Available from: https://www.healthit.gov/policy-researchers-implementers/clinical-decision-support-cds.

[12] Shojania KG, Jennings A, Mayhew A, Ramsay CR, Eccles MP, Grimshaw J. The effects of on-screen, point of care computer reminders on processes and outcomes of care. Cochrane Database Syst Rev 2009;3, CD001096.

[13] Joint Commission International Accreditation Standards for Hospitals. The joint commission; 2014. p. 23.

[14] Popovich D. 30-second head-to-toe tool in pediatric nursing: cultivating safety in handoff communication. Pediatr Nurs 2011;37:55–9.

[15] Davis J, Riesenberg LA, Mardis M, Donnelly J, Benningfield B, Youngstrom M, et al. Evaluating outcomes of electronic tools supporting physician shift-to-shift handoffs: a systematic review. J Grad Med Educ 2015;7:174–80.

[16] Li P, Ali S, Tang C, Ghali WA, Stelfox HT. Review of computerized physician handoff tools for improving the quality of patient care. J Hosp Med 2013;8:456–63.

[17] Savage SW, Schneider PJ, Pedersen CA. Utility of an online medication-error-reporting system. Am J Health Syst Pharm 2005;62:2265–70.

[18] Stavropoulou C, Doherty C, Tosey P. How effective are incident-reporting systems for improving patient safety? Milbank Q 2015;93:826–66.

[19] Daniel H, Sulmasy L. Physicians for the H and PPC of the AC of policy recommendations to guide the use of telemedicine in primary care settings: an american college of physicians position paper. Ann intern med 2015;163:787–9.

[20] Pade C, et al. An exploration of the categories associated with ICT project sustainability in rural areas of developing countries: a case study of the DWESA project. Proceedings of SAICSIT 2006;1:100–6.

[21] Batchelor S, Norrish P. Framework for the assessment of ICT pilot projects: beyond monitoring and evaluation to applied research. infoDev, World Bank Information and Development Program; 2009.

[22] Braa J, et al. Networks of action: sustainable health information systems across developing countries. MIS Q 2004;28(3):337–62.
[23] Krishna S, Walsham G. Implementing public information systems in developing countries: learning from a success story. Inf Technol Dev 2005;11(2):123–40.
[24] Sunden S, Wicander G. Information and communication technology applied for developing countries in rural context: towards a framework for analysing Factors influencing sustainable use; 2007. Fakultetenförekonomi, kommunikationoch IT, Sweden. Karlstad University Studies, 1403–8099; 2006:69 ISBN 91-7063-011-9.
[25] Heeks R. Health information systems: failure, success and local improvisations. Int J Med Inform 2006;75:125–37.
[26] Kimaro HC. Strategies for developing human resource capacity to support sustainability of ICT based health information systems: a case study from Tanzania. Electron J Inf Syst Dev Ctries 2006;26(2):1–23.
[27] Hebert R. Health informatics—where to start? national E-health options for developing countries. Inf Tech Dev Ctries 2008;18(1). IFIP WG 9.4.
[28] Kimaro HC, Nhampossa JL. The challenges of sustainability of health information systems (HIS) in developing countries: comparative case studies of Mozambique and Tanzania. In: The 12th European conference on information systems (ECIS)—the European IS profession in the global networking environment; 2004.
[29] Anon. How healthy are rural Canadians? An assessment of their health status and health determinants. Ottawa: Canadian Institute for Health Information; 2006, ISBN:978-1-55392-881-2.
[30] Stephens S. Gap in life expectancy between rural and urban residents is growing. Center for Advancing Health; 2021.
[31] Anon. Rural, regional, and remote health: Indicators of health. Australian Institute of Health and Welfare; 2005, ISBN:9781740244671.
[32] Then J. Analysis of urban-rural population dynamics of China: a multiregional life table approach. Environ Plan 1993;25(2):245–53.
[33] Health workers needed: poor left without care in Africa's rural areas. The World Bank. 2008.
[34] Bring health care services to rural Africa. The Atlantic Philanthropies. 2012. https://www.atlanticphilanthropies.org.
[35] Health. African solutions to African problems. 2013. https://www.crisisgroup.org.
[36] Faux SG, Eager K, Cameron ID, Poulos CJ. COVID -19: planning for the aftermath to manage the aftershocks. MJA 2020;29.
[37] Bettger JP, Thoumi A, Marquevich V, De Groote W, Battistella LR, Imamura M, Ramos VD, Wang N, Dreinhoefer KE, Mangar A, Ghandi DB. COVID-19: maintaining essential rehabilitation services across the care continuum. BMJ Glob Health 2020;5(5), e002670.
[38] Care D. Rehabilitation and physiotherapy in times of pandemic. Available from https://www.dycare.com/products/rehabilitation-and-physiotherapy-in-times-of-pandemic/.
[39] Achenbach SJ. Telemedicine: benefits, challenges, and its great potential. Health Law Policy Brief 2020;14(1). Available at: https://digitalcommons.wcl.american.edu/hlp/vol14/iss1/2.
[40] Shaw DK. Overview of telehealth and its application to cardiopulmonary physical therapy. Cardiopulm Phys Ther J 2009;20(2):13–8. https://doi.org/10.1097/01823246-200920020-00003.
[41] Miller EA. Solving the disjuncture between research and practice: telehealth trends in the 21st century. Health Policy 2007;82(2):133–41.

[42] Dinesen B, Nonnecke B, Lindeman D, Toft E, Kidholm K, Jethwani K. Personalised telehealth in the future: a global research agenda. J Med Internet Res 2016;18(3), e53.
[43] Digital Physical Therapy Task Force. Report of the WCPT/INPTRA digital physical therapy practice task force. In: World confederation for physical therapy; 2019. 24 p. Report No. 7. Available from https://www.wcpt.org/sites/wcpt.org/files/files/wcptnews/REPORT%20OF%20THE%20WCPTINPTRA%20DIGITAL%20PHYSICAL%20THERAPY%20PRACTICE%20TASK%20FORCE.pdf.
[44] Masson M. Benefits of TED talks. Can Fam Physician 2014;60(12):1080.
[45] 2010 Opportunities and developments | Report on the second global survey on eHealth | Global Observatory for eHealth series - Volume 2: Telemedicine. 2011.
[46] What is Telemedicine?. Washington, D.C. American Telemedicine Association. 2013.
[47] Yamin R, Bathish J, Berman Y, Isa A, Zeidman A, et al. In hospital physiotherapy treatment for Covid-19 patients—management and clinical practice. Arch Pulmonol. Respir Care 2020;6(1):017–20. https://doi.org/10.17352/aprc.000044.
[48] Zundel KM. Telemedicine: history, applications, and impact on librarianship. Bull Med Libr Assoc 1996;84(1):71–9.
[49] Gershon-Cohen J, Cooley AG. Telediagnosis. Radiology 1950;55:582–7.
[50] Allen A. Teleradiology I: introduction. Telemed Today 1996;4(1):24.
[51] Allen A, Allen D. Teleradiology 1994. Telemed Today 1994;2(3):14–23.
[52] Perednia DA, Allen A. Telemedicine technology and clinical applications. JAMA 1995;273(6):483–7.
[53] Jutra A. Teleroentgen diagnosis by means of videotape recording. Am J Roentgenol 1959;82:1099–102.
[54] Davis DA, Thornton W, Grosskreutz DC, et al. Radio telemetry in patient monitoring. Anesthesiology 1961;22(6):1010–3.
[55] Hirschman JC, Baker TJ, Schiff AF. Transoceanic radio transmission of electrocardiograms. Dis Chest 1967;52(2):186–90.
[56] Bird KY. Cardiopulmonary Frontiers: quality health Care via interactive television. Chest 1972;61:204–5.
[57] Puskin DH, Brink LH, Mintzer CL, et al. Joint federal initiative for creating a telemedicine evaluation framework. Telemed J 1995;393–7. Letter to the Editor.
[58] Vockley M. The rise of telehealth: 'triple aim,' innovative technology, and popular demand are spearheading new models of health and wellness care. Biomed Instrum Technol 2015;49(5):306–20.
[59] Kvedar J, Coye MJ, Everett W. Connected health: a review of technologies and strategies to improve patient care with telemedicine and telehealth. Health Aff 2014;33(2):194–9.
[60] Hirani SP, Rixon L, Beynon M, Cartwright M, Cleanthous S, Selva A, Sanders C, Newman SP. Quantifying beliefs regarding telehealth: development of the whole systems demonstrator service user technology acceptability questionnaire. J Telemed Telecare 2017;23(4):460–9.
[61] Ryan Haight Act will Require Tighter Restrictions on Internet Pharmacies. 2010. https://www.govtech.com/health/ryan-haight-act-will.html.
[62] Langarizadeh M, Moghbeli F. Applying naive bayesian networks to disease prediction: a systematic review. Acta Inform Med 2016;24(5):364.
[63] Langarizadeh M, Saeedi M, Far M. Hoseinpour M. Predicting premature birth in pregnant women via assisted reproductive technologies using neural network. J Health Adm 2016;18(62).

[64] Dye C, Reeder JC, Terry RF. Research for universal health coverage. American Association for the Advancement of Science; 2013.
[65] White-Williams C, Oetjen D. An ethical analysis of telemedicine: implications for future research. Int J Telemed Clin Pract 2015;1(1):4–16.
[66] Tarzian AJ. Force ACCUT health care ethics consultation: an update on core competencies and emerging standards from the American Society for Bioethics and Humanities' Core competencies update task force. Am J Bioeth 2013;13(2):3–13.
[67] Dombo EA, Kays L, Weller K. Clinical social work practice and technology: personal, practical, regulatory, and ethical considerations for the twenty-first century. Soc Work Health Care 2014;53(9):900–19.
[68] Parsons TD. Telemedicine, mobile, and internet-based neurocognitive assessment. Clin Neuropsychol Tech: Springer 2016;99–111.

CHAPTER

Global rural health-care outlooks

5

As compared to urban areas, the present health status of the rural community is extremely poor. For example, in South Africa, infant mortality rate (IMR) in rural areas are 1.6 times higher than that in urban areas. Even in developing countries like India, the mortality rate in rural areas is 1.5 times higher than that in urban areas.

Keeping in view the increasing rate of poverty, low health status, and high burden of disease in rural areas, it is high time to improve the present status of poor health conditions. The WHO International Development Programme has highlighted this, with specific objectives for policies and actions that promote sustainable livelihood including access to land, resources, and market, as well as better education, health, and opportunities for rural people. World Health Day is acknowledged by various governments and nongovernmental organizations with interests in public health issues.

There are other health awareness days, also, being officially sponsored by WHO, and those days include World Tuberculosis Day, World Immunization Week, World Malaria Day, World No Tobacco Day, World AIDS Day, World Blood Donor Day, World Chagas Diseases Day, World Patient Safety Day, World Antimicrobial Awareness Week, and World Hepatitis Day.

In 1971, in order to promote health care and diseases-prevention, the US Department of Health and Human Services initiated a program to promote health care and diseases-prevention. This was in response to an emerging consensus among scientists and health authorities to prevent disease, on a priority basis. In this connection, the first issue contained "a report announcing goals for a 10-year plan to reduce the controllable health risk. In its section on nutrition, the report recommended diets with fewer calories; less saturated fat; cholesterol, salt, and sugar; relatively more complex carbohydrate, fish and poultry, and less red meat." In addition, it was also advised to "be wary of processed foods." The goals were subsequently updated for Healthy People 2000 and Healthy People 2020. The overall goals of healthy people 20 are to provide high-quality, longer lives free of preventable diseases, disability, injury, and premature death, achieve health equity, eliminate disparities, and improve the health of all groups which would lead to promote good health for all and to bring sustainable healthy life.

5.1 Health-care access in rural communities

Access to health-care services is a decisive factor for bringing sustainability in community health, even under adverse natural conditions. A 1993 National Academies

Healthcare Strategies and Planning for Social Inclusion and Development. https://doi.org/10.1016/B978-0-323-90446-9.00005-8

report says, Access to Health in America, defined access as the "timely use of personal health services to achieve the best possible health outcomes." The Rural Policy Research Institute (RUPRI) health panel included accessibility based on five basic principles which include affordability, accessibility, community health, high-quality care, and patient centeredness [1].

The possibility of success in developing rural health care is primarily based on a clear understanding of the access to health care by researchers, policy makers, NGOs (nongovernment organizations), and providers. But, the rural communities confront certain barriers to avail health-care facilities such as even an adequate supply of health-care services being provided by the government. A variety of elements contribute to these problems in rural areas, including population, economic stagnation, shortage of physicians and other health-care professionals, a disproportionate number of elderly, poor, and underinsured residents, and high rates of chronic illness.

In order to find harmony in health-care services at the rural community level, many international organizations have been working to resolve various issues related to the implementation of action plans on health-care services for rural communities in underdeveloped nations. In this connection, the United Nations Sustainable Development Goal (UN SDGs) is unique having targeted to eradicate poverty and develop sustainable health care, especially in rural areas and undeveloped nations by the end of 2030. Under the health goal (SDG3) it has been ensured universal health coverage (UHC) and access to quality health care. The SDG 3 ensures health and well-being for all, including a bold commitment to end the epidemics, AIDs, tuberculosis, malaria, and other communicable diseases by 2030 (Fig. 5.1). It is also targeted to achieve universal health coverage and provide access to safe and effective medicines and vaccines for all.

Sustainable Development Goal 3 covers all major health priorities, including infant mortality; mental health; reproductive health; child care; communicable, noncommunicable, and environmental diseases; universal health coverage; and access to safe and effective medicines for common people.

FIG. 5.1

Symbolic expression of SDG3.

5.1.1 Rural area

In general, a rural area or countryside is a geographic area that is located outside towns and cities. "Typical rural areas have a low population density and small settlements with fewer than 2500 residents or areas designated as rural can have population densities as high as 999 per square mile or as low as 1 person per square mile" [2]. The Health Resources and Human Services defines the word rural as encompassing "all population, housing, and territory not included within an urban area whatever is not urban is considered rural." There is no international standard for defining rural areas, and standards may vary even within an individual country.

5.1.2 Significance of accessibility

Implementation of well-planned and effective health-care services is important for boosting sustainable health in rural communities by eradicating and preventing contagious and noncontagious diseases. Access to health services means "the timely use of personal health services to achieve the best possible health outcomes" [3].

The four major steps for effective implementation of health-care services are [4]: (i) coverage: it includes providing entry into the health-care system because uninsured people are less likely to avail health care and more likely to have poor health status; (ii) services: related to prevention of diseases in communities; (iii) timeliness: providing timely health-care services; and (iv) workforce: to provide well-trained health-care workforces to assist doctors and campaign for primary knowledge on health care. But, implementation for such services becomes a failure due to lack of budget to meet high-cost health-care services, inadequate or no insurance coverage, lack of availability of services, and lack of culturally competent care. But, it has been observed that access to care often varies based on race, ethnicity, socioeconomic status, age, sex, disability status, sexual orientation, gender identity, and residential location [5, 6].

5.1.3 Health-care access in rural communities

In rural areas, the access to health services is comparatively lesser than their urban counterparts. Limited number of medical practitioners, poor primary health center, less developed communication services, nonavailability of primary education, lack of health-care workers, nonavailability of proper network services are some of the important factors responsible for chronic health problems in rural areas.

Access to health-care services plays a critical role in developing good health (Fig. 5.2).

Still, rural residents all over the world confront a variety of problems in accessing health-care services being provided by the concerned government. In 2014, the Rural Policy Research Institute (RUPRI) Health Panel envisions rural health care that is affordable and accessible for rural residents through a sustainable health system that is responsible for delivering high-quality and high-value services.

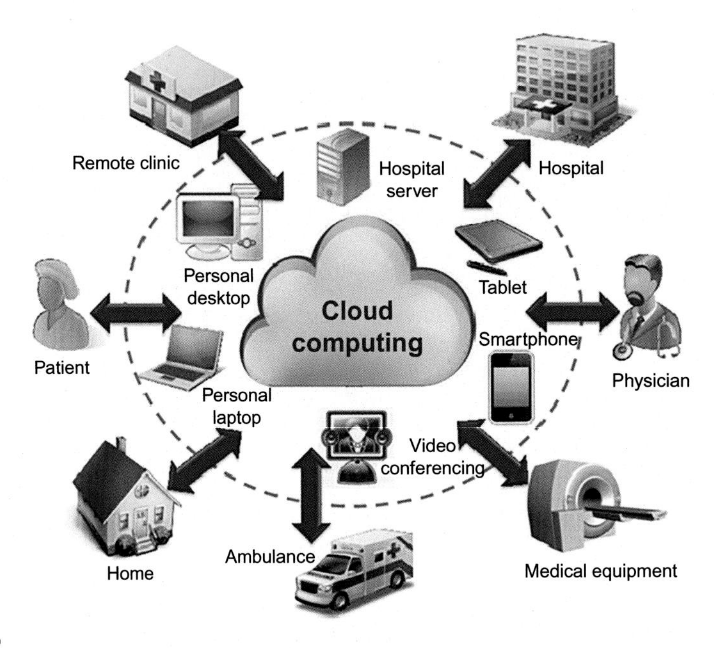

FIG. 5.2

Common application to virtual care delivery through cloud computing.

The government of the concerned state should provide necessary opportunities to an individual so that he or she can able to conveniently and confidently access services such as primary care, dental care, behavioral health, emergency access to health care. But, it has been observed that rural residents confront many hurdles to avail access to health care.

5.2 Social determinants on health

5.2.1 History

In the early 2000s, the WHO facilitated the academic and political work on social determinants in order to bring harmony in health-care services to remove disparities from a global perspective point of view. In 2003, WHO, Europe suggested that the social determinant of health include the social gradient, social exclusion, early life, unemployment, social support, addiction, food, and transport.

In 2005, in order to give social justice, The Commission on Social Determinants of Health was set up to promote health equity and to foster a global movement to achieve it. The Commission calls on the WHO and all governments to go ahead with action on the social determinants, civil society, WHO, and other global organizations to join hands to improve the lives of the World's citizens.

In 2008, the WHO Commission on Social Determinants of Health published a report entitled "Closing the Gap in a Generation" on how health inequity could be remedied and what action could combat factors that exacerbated injustices [7–10].

In 2010, The Affordable Care Act (ACA), also known as Obamacare, was signed into law in March 2010. It was designed to extend health coverage to millions of uninsured Americans by legally requiring them to buy health insurance. The comprehensive health-care reform law was enacted in March 2010 (known as ACA, PPACA, or Obamacare) [11, 12].

The law has three primary goals:

- Make affordable health insurance available to more people. The law provides consumers with subsidies (premium tax credit) that lower costs for householders within incomes between 100% and 400% of the federal poverty level.
- Expand the medical program to cover all adults with income below 138% of the federal poverty level.
- Support innovative medical care delivery methods designed to lower the costs of health care generally.

In 2011, the WHO convened a global conference in Rio de Janeiro, Brazil, to build support for the implementation of action on social determinants of health. The conference brought together over 1000 participants 125 Member States and a diverse group of stakeholders.

5.2.2 Factors for social determinants

The social determinants of health (SDH) are the conditions of habitable life into which people are born and spend their life under the circumstances of a variety of factors like socioeconomic conditions and the wider set of forces and systems that shape the conditions of their daily life (Fig. 5.3).

Virtually, the overall contribution of medical health care is relatively small as compared to the outcome of social, behavioral, and physical environment factors. The contribution of nongovernmental organization (NGO) and other international bodies is enormous due to their close association with a wide range of services like access to safe, stable housing, nutritious food, counseling services, recreation programs, transportation, and advocacy.

These forces and systems include economic policies and systems, development agendas, social norms, social policies, and political systems. As compared to the living habit of urbanized localities, the SDH is under great disparity in health status, especially in developing and underdeveloped countries. In general, the major problems related to SDH can be categorized into five broad groups: genetics, behavior, environmental and physical influences, medical care, and social factors.

Both the urbanized locality and rural area people are victimized with SDH. Irrespective of the nature of habitable life, lifestyles, and the conditions in which people live and work strongly influence their health and longevity. Medical care can cure the problems of people suffering from disease, but socioeconomic conditions, and suppressive professional working style may cause serious health disorders which cannot be guaranteed to be cured. On the pathway of life, health is resolved on the basis of complex interactions between social and economic factors, the physical

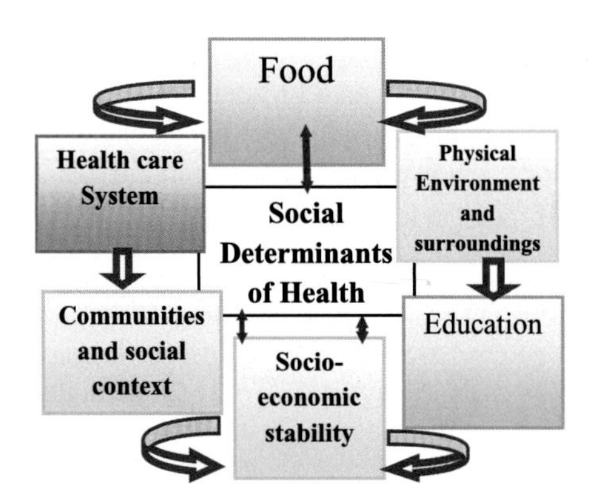

FIG. 5.3

Graphic presentation of social determinants of health associated with various factors.

environmental, and individual behavior. They do not exist in isolation from each other. Followings are few important aspects responsible for SDH, may be directly or indirectly.

Income and social status

The social status in health means that health inequities affect everyone. The social status in health is a term used to describe the phenomenon whereby people who are less advantaged in terms of socioeconomic position have worse health than those who are more advantaged (Fig. 5.4A and B).

The maintenance of health care mainly depends on the gradient of socioeconomic conditions of an individual. This is a global phenomenon, seen in lower-, middle-, and high-income countries. Poor social and economic are chronic factors that affect health throughout life and are responsible for more than twice the risk of serious illness and premature death of those near the top. A continuous social gradient is prevailing from top to bottom and responsible for more detritus unhealthy life for the people at the bottom.

Social support networks

A social support network is made up of friends, family, and peers. Social support is different from a support group, which is generally a structured meeting run by a lay leader or mental health professional. Social support networks may not be directly helpful in explaining health outcomes for illnesses over which the individual has little control. But in case of any hazardous community health problems, the social support networks play a significant role in producing better health (Fig. 5.5).

Social relations and Supportive networks boosts the concept for "thinking for each other," and establish healthy social interaction which can be helpful in developing "health for all" feelings among the people. Social networks are all-encompassing sets for the link by which an individual can understand and establish his or her identities, which can be ultimately beneficial to create a favorable attitude for developing a concept on "health for all" [13, 14].

FIG. 5.4

Photo of habitable life of (A) poor people (site unauthorized colony nearby railway track, photo taken by authors), and (B) urbanized life.

Physical environment

Physical determinants include natural environment, such as green space (e.g., tree and grass) or weather (e.g., climate change); and built environments, such as buildings, sidewalks, bike lanes, and roads. It has been observed that the determinant of health is also significantly related to the social environment encompassing socioeconomic condition, political and cultural spheres, and the possibility of a sustainable

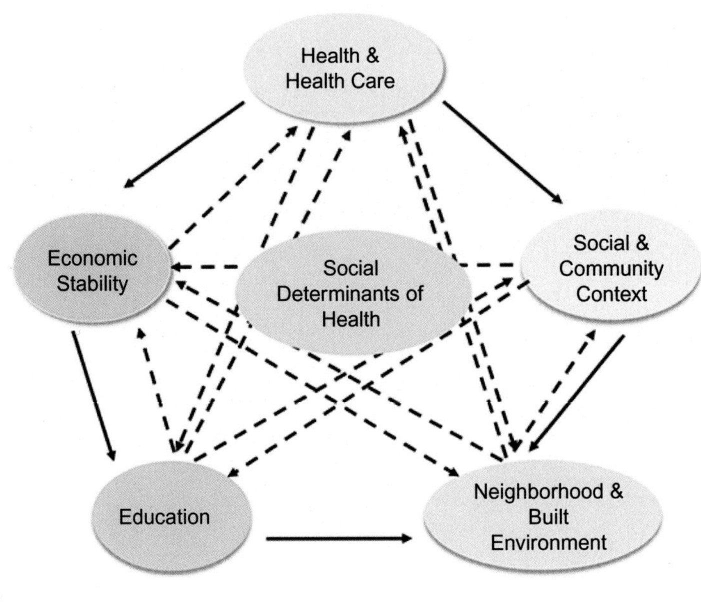

FIG. 5.5

Social support network system for social determinants of health.

environment. The social environment also includes moral sentiments defining good and bad thoughts, feelings and conduct, ideologies including religious and secular beliefs. The social environment provides (Fig. 5.6) social and economic safety, cohesion in a community, and so on. The best example of a social environment is the state of the composition of the family. In general, the social environment consists of the customs and traditions of the society in a habitable life that is supposed to continue smoothly. It includes the standard of living in society.

Healthy people 2020 highlights the significance of addressing the social determinants of health through adaptable social and physical environments, which can promote good health for all. This emphasis is shared by the World Health Organization, whose Commission on Social Determinants of Health in 2008 published the report, "Closing the gap in a generation: Health equity through action on the social determinants of health."

Education and literacy

Opportunity for proper education, especially primary education, helps the rural people in reducing health disparities, improving quality of life, and addressing social determinants of health (SDOH). In this connection, early childhood education programs aim to lay a foundation of health by building the social and emotional skills of young children. In addition, early childhood education programs can have profound long-term and far-reaching impacts on health and well-being over a lifespan. It has been observed that providing early childhood education and positive developmental experiences have been shown to improve overall educational attainment, increase potential future earning, reduce crime rates, and improve health outcomes. So, in the developed countries the Community Services Task Force (community health-care manager) recommends the implementation of educational interventions, such as center-based early childhood education (ECE) to reduce health inequities and to support children who live in communities with an uneven distribution of SDOH, including low-income populations. Adults with higher education can also guide and help people lead a better healthy life.

Social exclusion

Social exclusion has a significant relation with health and the area of primary health care in particular (Fig. 5.7). Social exclusion can be defined in many ways. But, mostly it is described as the propellers, especially rural women's confrontation to adjust themselves when removed from the mainstream of life [15–17].

Basically, the phrase social exclusion originated in 1970s France to express adversity faced by a group of citizens who were not provided by the state social security net [18]. In the later stage, the European Commission later introduced the term social exclusion into discussion alongside the term "poverty" for many programs and initiatives from the early 1990s [19–21].

The United Nations Sustainable Development Goals (SDGs), and goal number three in particular, which is related to health and well-being across the life course. The main target of this goal is the effective management of conditions such as HIV and substance use and the introduction of universal health coverage among other targets. The main idea behind this is that improving the health status of such socially excluded groups may improve the health of the population as a whole.

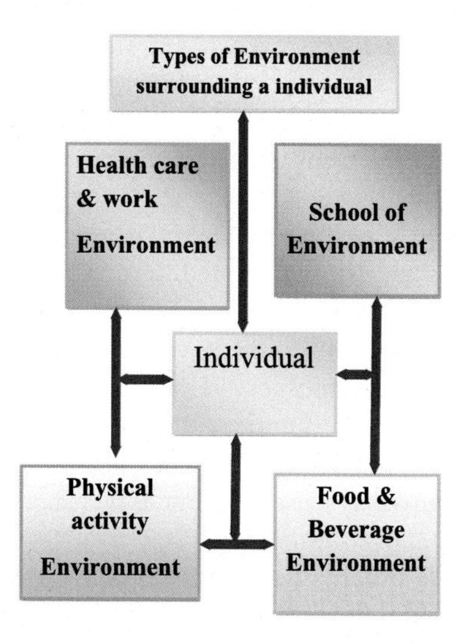

FIG. 5.6

Nature of physical environment surrounding an individual.

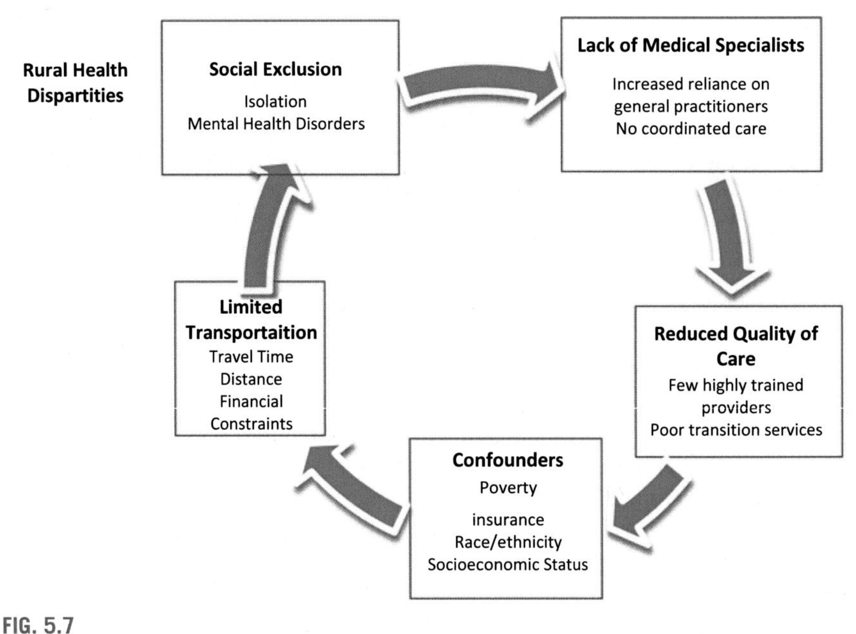

FIG. 5.7

Social exclusion and health care.

Gender equity

Gender plays a crucial role as a determinant of health service. But, gender is increasingly important in understanding how women and men experience and respond to health promotion programs and interventions and their outcomes. Gender, as a determinant of health, refers to interrelated dimensions of biological difference, psychological difference, and social experience. Besides the biological differences, gender roles, norms, and behavior influence how women, men, girls, and boys access health services and how the health system responds to their different needs.

Gender inequality may bring unrepeatable damages to the physical and mental health of millions of girls and women around the world. The problem of health inequality is mainly caused by the ignorance of the government on women's rights to health. At the primary education level, the government should make primary education compulsory for all people, irrespective of gender. Due to the involvement of large number of people and the magnitude of the problems with health-care disparity, one of the most direct and potent ways to reduce health inequalities and ensure effective use of health resources is to strictly implement health-care services at community level.

But, despite worldwide progress, gender inequality is still prevailing in society. Far too many girls, especially those from the poorest families, still face discrimination with respect to basic education, marriage and pregnancy, sexual violence, and unrecognized domestic work.

Healthy child development

Health child development means that children of all abilities, including those with special health care needs, are able to grow up where their social, emotional, and educational needs are met.

The WHO wants to focus on 10 important facts on early child development as a social determinant of health.

Fact 1

Take care of children from the beginning of gestation to 8 years of age in order to smoothen their physical, social, emotional, and language/cognitive development.

Fact 2

Stable, responsive, and nurturing are the three key issues to be taken care in order to provide safe, supportive environments and appropriate nutrition. Exclusive breastfeeding in the first 6 months, timely vaccination, as prescribed by the physicians, and appropriate nutrition are also equally important during the development phase of children.

Fact 3

It aims to lay a strong foundation in children in order to maintain sound physical and mental health, enhancing the ability to adapt to change and build the capacity for resilience against adverse circumstances.

Fact 4

Promote and ensure stable, caring, interactive relationships between adults and children so that healthy brains would develop in young children.

Fact 5

Safe, cohesive, child-centered neighborhood, communities, and villages matter for early child development.

Fact 6

The main target is to provide high-quality early childhood care and primary education, which can be helpful in improving children's chances for success in the later phase of life. This would be helpful in improving the lives of the most disadvantaged and vulnerable children and their respective families.

Fact 7

The government and global communities need to have a support system to improve the conditions of families to nurture their children by addressing economic security, flexible work, information and support, health, and quality childcare needs.

Fact 8

Promote and ensure early childhood development with a comparable approach throughout the world that will provide a way for societies to judge their success in early child development at their respective community level.

Fact 9

Develop efficient survival indicators for child mortality rate in order to have an understanding of child survival and child development without any conflict. This would be helpful to minimize the number of children who fail to achieve their full development potential.

Fact 10

The objective is to avoid the children from exposure to physical, biological, and chemical hazardous risks in their early developmental stage. This would save children from problems such as physical trauma, biological diseases, and chemical hazards.

Employment

Population health is critically related to employment and working conditions. It has been observed that education level, age, marital status, household labor force, and total government grants are significant determinants of employment status. Employment is one of the most shaping factors leading to safe and secure life for good health. The Employment Conditions Knowledge Network (EMCONET) is helpful in developing models and measures to clarify how different types of jobs, conditions of underemployment, and the threat of becoming unemployed affect worker's health and finds ways and means to improve working conditions and provides fair access to employment and other dimensions of decent work.

Personal health practices and coping skills

Personal health practices are keys to prevent diseases and promote self-care. Effective coping skills enable people to be self-reliant, solve problems, and make choices that enhance health. At the individual level, one can prevent diseases, promote self-care, and make choices that enhance health. But, it does not necessarily mean that lifestyle includes individual choices and the influence of social, economic, and environmental factors on the decisions people make for their health. Personal life "choices" are influenced by the socioeconomic environments in which people live.

Biology and genetic endowment

The basic structural get up of the human body is the most critical aspect for the determinant status of health. In addition, inherited genetic factors are also responsible for an individual response that can affect status throughout the lifespan.

Health services

Effective implementations of health services promote health and prevent diseases that influence health. The most important health services include protecting water supplies, primary education related to health care, and health counseling have an impact on overall health.

Culture

Sociocultural determinants are defined as "community's or society's attitudes, beliefs, and values related to health behavior." So, the community or social customs, traditions, and beliefs affect health. A person's cultural background has an influence on their beliefs, behaviors, perceptions, emotions, language, diet, body image, and attitudes to illness, pain, and misfortune. All of these factors can influence health and the use of health-care services.

5.3 Strategies to improve rural health

Health-care quality is a variable issue in different countries. Rural people of developing countries face more hurdles in maintaining their health quality. In order to improve health quality, rural providers, like their urban counterparts, must follow a comprehensive approach to quality improvement. In this connection, it is advised that the rural health care providers must have to follow practice guidelines and computer-aided decision support, standardized performance measures, data feedback capabilities, and quality improvement processes and resources. It is necessary to improve or bring amendments to serious shortcomings in the quality of health care, in both rural and urban communities. A single solution in this regard may not be helpful, as rural and urban areas vary in their beliefs, customs, practices, and social behavior.

In connection with the quality challenge, the rural people must develop their strengths in order to overcome problems such as scarcity in delivering health-care services in rural locality, the lack of resources, and providers' availability. A variety of elements contribute to these problems in rural areas, including a declining population, economic stagnation, shortage of physicians and other health-care professionals, a disproportionate number of elderly, poor, and underinsured residents, and high rates of chronic illness.

The success in quality health-care development mainly depends on effort in the leadership program, cultural change in organizations, and human resources, and technical support. The quality of rural infrastructure can be improved by improvizing the local communication system, renovation of housing patterns, and environmental upgrading. It is also necessary to bring value-added touch to classic rural heritage. In this regard, rural communities have long struggled to maintain access to quality care services [22–25].

For health-care development, rural communities need access to clinical knowledge and the tools needed to implement this knowledge to practice.

The rural health-care providers can pursue multifaceted strategies to improve health care: (i) availability, (ii) accessibility, and (iii) affordability.

5.3.1 Availability

Availability means the sufficient supply and appropriate stock of health workers, with the competencies and skill mix to match the health needs of the individual, especially belonging to rural locality.

Access to the availability of comprehensive, quality health-care services is also important for promoting and maintaining health, preventing and managing disease, reducing unnecessary disability and premature death, and achieving health equity for all. Availability of timely health care is the most important for maintaining good health in rural areas. This could only be possible by team-based care and timeliness of care. Telehealth is a potential technology but has drawbacks such as needing to travel to a medical practitioner to use a secure telehealth service. This would be helpful in meeting the need of the shortage of specialists in a rural area, but there are challenges linked to telehealth in rural areas such as regulatory and licensing restrictions. For example, telehealth providers are often required to be in the same state as the patient.

5.3.2 Accessibility

Accessibility means equitable distribution of health workers taking into account the demographic composition, rural-urban mix, and underserved areas or population.

The health care workers should have sufficient access, necessary and appropriate health-care providers to serve the rural communities. Even when an adequate supply of health-care services exists in the community, there are other factors to consider in terms of health-care access. For example, to have good health-care access, a rural resident:

- must have the potential to pay health insurance that is accepted by the provider;
- must have means to reach and use services, such as transportation to services that may be located at a distance, and the ability to take paid time off of work to use such services;
- should possess confidence in their ability to communicate with health-care provider;
- should trust the provider without possessing any sort of reservation in expressing the health problem.

Availability and accessibility are the complementary factors for health workers in the success of good health-care services. In addition, the acceptability of an individual is very important. Without acceptability, the health services might not be used, when the quality of the health workforce is inadequate, improvements in health outcomes will not be satisfactory.

5.3.3 Affordability

Affordable basic health care is a critical issue for a rural individual to meet health-care services given by the provider or health manager. This is because expenditure around the world continued to rise as a percentage of gross domestic product. The treatment of some life-threatening diseases by biological drugs is supposed to be highly expensive. So, it is necessary that the government should have a provision of some sort of subsidy or insurance policy for the rural people to meet the expenses in treatment, at the time of need.

5.4 Organization working in global health

5.4.1 Multilateral agencies

Generally, there are three groups of an international organizations working around the world to have look after on health at the time of crisis due to natural disaster or any man-made misshapen. The international health care organizations include (i) multilateral organization, (ii) bilateral organization, and (iii) nongovernment organizations (NGO). These organizations manage to get funds from multiple government and nongovernmental sources and distributed among many different countries.

5.4.2 Multilateral organization

World Health Organization

The World Health Organization (WHO) was established soon after World War II as a multilateral health organization to unite countries in the common goals of handling diseases for a healthy world. In 1977, the World Health Assembly decided that the social target of governments, and WHO should be the attainment by all the people of the world by 2000 of a level of health that would permit them to lead a social and economically productive life. The organization recruits experts with a wide array of expertise, including medical doctors, researchers, epidemiologists, administrative staff, statisticians, economists, and other related fields to operate a wide range of programs and projects worldwide. In addition, WHO is having a target to prolong life and improving quality of life through organized efforts and informed choices of society, organization, communities, and individuals. Analyzing the determinants of health of a population and the threats it faces. As defined by WHO the public health can be as small as a handful of people or as large as a village or entire city; in case of a pandemic, it may encompass several countries. The concept of health takes into account physical, psychological, and social well-being.

World Bank

The World Bank is a leading institution for investments in health and development. Currently, the World Bank funds more than $1 billion annually to improve health, nutrition, and population in developing countries. Moreover, it is one of the world's largest external funders of the fight against HIV/AIDS, with current commitments of more than $1.3 billion. The organization offers a variety of career and employment opportunities, professional development, and internship opportunities.

United Nation International Children's Fund

United Nations International Children's Emergency Fund (UNICEF) is a United Nations agency that provides aid for children welfare all over the world, from the humanitarian point of view [26].

The agency is among the most widely recognizable social welfare organizations in the world, with a presence in 192 counties and territories. The main activities of UNICEF include providing immunizations and disease prevention, administering treatment for children and mothers with HIV, enhancing childhood and maternal nutrition, improving sanitation, promoting education, and providing emergency relief in response to disasters.

Bilateral Agencies

Bilateral Agencies receive funding from the government in home countries and operate directly between two counties. Funds flow from official sources directly to official sources in the recipient country.

United States Agency for International Development (USAID)

The USAID is one of the largest bilateral agencies involved in global health efforts. The USAID is an independent agency of the United States federal government that is primarily responsible for administering civilians foreign aid and development assistance. USAID provides funding for and supports global health initiatives in areas such as emerging pandemic threats, family planning, HIV and AIDS, health systems strengthening, malaria and child health, neglected tropical diseases, nutrition, and tuberculosis (TB).

Center for Diseases Control and Prevention

Center for Disease Control and Prevention (CDC) is a national public health institute in the United State. It is a United States federal agency, under the Department of Health and Human Services. The main target of the CDC is developing and applying disease control and prevention. It especially focuses attention on infectious disease, foodborne pathogens, environmental health, occupational safety and health, health promotion, injury prevention, and educational activities in improving the health of the citizens of the United States.

Nongovernmental organization

Nongovernmental organizations (NGOs) are generally defined as nonprofits entities independent of governmental influence, although they receive government funding. The NGOs are task-oriented and are actively engaged in various issues of the communities such as health, human rights, or the environment. Examples of nongovernmental organizations:

Medicines Sans Frontiers (MSF)

It is an organization having a target to help people worldwide related to various issues on health. It delivers emergency medical aid to people impacted by conflict, epidemics, disasters, or lack of access to care.

CARE International. The main target of this organization is to eradicate global poverty. The organization places special emphasis on working with and empowering women as a way to help whole families and communities escape poverty. The organization also financially support the people involved in advocacy, education, maternal health, HIV and AIDS, and food security.

Population Services International. Population Services International (PSI) is a nonprofit organization with programs targeting malaria, child survival, HIV, and reproductive health. PSI also provides products, clinical services, and behavior change communications for the health of people in high-need populations.

5.5 Universal health coverage (UHC)

As per the WHO concept, universal health coverage means that "all people have access to the health services they need, when and where they need them, without financial hardship." It includes the full range of essential health services from health promotion to prevention, treatment, rehabilitation, and palliative care.

At present, about 50% of health care needy people are left unseen from health services. At the global level, about 100 million people are under extreme poverty each year. Health for all could only be possible when both the individual and community will have access to timely quality health care. Universal health coverage should be based on strong, people-centered primary health care. The good health system is rooted in the communities they served. They focus not only on preventing and treating disease and illness but also on helping to improve well-being and quality of life. UHC allows countries to make the most of their strongest asset: human capital. The countries' economic growth mainly depends on UHC. Without good health, the overall activities of life get stuck.

5.5.1 Toward universal health coverage

Global health strategies established Universal Health Coverage (UHC) Day on 12 December to bring awareness among the people on health for all. Health is also an essential part of the Sustainable Development Goals (SDGs). For example, the SDG 3.8 target aims to "Achieve universal health coverage, including financial risk protection, access to quality essential health coverage, including financial risk protection, access to quality, and affordable essential medicines and vaccines for all." In addition, SDH 1, which calls to "end poverty in all its forms everywhere." In recent years, the UHC movement has gained global momentum, with the first-ever UN high-level meeting on UHC held in September 2019. Subsequently, in January 2020, the second UHC Forum was held in Bangkok, aiming to enhance political momentum on UHC in the international forum.

Moving toward UHC requires strengthening the health system in all countries. It is difficult for common people to meet the cost of health services out of pocket. Generally, the poor rural people are unable to obtain many of the services they need, and the rich may be exposed to financial hardship in the event of severe or long-term illness. Pooling funds from compulsory funding sources (such as mandatory insurance contributions) can spread the financial risks of illness across a population. To achieve successfully, investment in quality health care around the world is an essential fact, as managing well-trained workforce is highly expensive. Besides this, good governance, an efficient supply chain management system for health aids and medicines, and availability of modern health technologies, and well-functioning health information systems are other critical elements for achieving UHC around the world.

5.5.2 Primary health care vs UHC

Basically, primary health care is services for good health on the needs and circumstances of individuals, families, and communities. It includes comprehensive and interrelated physical, mental, and social health and well-being. It is supposed to be a provision for whole-person care for health needs throughout life, not just treating a set of special diseases. Primary health care looks after comprehensive care covering the treatment and prevention of diseases, and attempt to the rehabilitation of the individual with everyday environment. WHO has defined primary health care as: "ensuring people's health problems are addresses through compressive primitive, protective, curative, rehabilitative, and palliative care throughout the life course, strategically prioritizing key system functions aimed at individuals and families and population at the central elements of integrated service delivery across all levels of care; Systematically addressing the broader determinants of health (including social, economic, environmental, as well as people's characteristics and behaviors) through evidence-informed public policies and behaviors) through evidence-informed public policies and actions across all sectors; and empowering individuals, families, and communities to optimize their health, as advocates for policies that promote and protect health and well being, as co-developers of health and social services through their participation, and as self-cares and care-givers to others."

Universal health care can be achieved by the successful implementation of primary health care services. Universal health care emphasizes both the quality and successful operation (proper funding, distribution pattern, management) of UHC policies all over the world. The health-care services include reframing health services of both indoor and outdoor patients in a harmonic pattern. WHO also defined health services which, including traditional and complementary medicine services, to fulfill the needs and expectation of people and bringing the feeling of confidence to take a more active role in their health and health system.

5.5.3 Funding model

In order to provide better health coverage, the government worldwide are endeavoring to work out the best way to meet citizen's expectation. In this regard, the WHO Member States have set themselves the target of developing their health financing systems in ways that are capable of ensuring a balanced package for universal coverage within the purview of their financial budget.

Generally, UHC gets funding for organizing and implementation of decisions from tax revenue (as Portugal, India, Spain, Denmark, and Sweden). Some nations, such as Germany, France, and Japan, employ a multipayer system in which health care is funded by private and public contributions. However, much of the nongovernment funding comes from contributions from employers and employees to regulated nonprofit sickness funds. All types of contributions as stated above are compulsory and defined according to law.

5.5.4 WHO's role

As part of Sustainable Development Goals, United Nations member states have agreed to work toward worldwide universal health coverage by 2030. UHC program is based on the 1948 WHO Constitution which declares health a fundamental human right and committed to ensuring the highest attainable level of health for all. WHO has been in the process of supporting countries to develop their health systems to move toward and sustain UHC, and monitor progress. WHO's work is aligned with SDG target 3.8, which cover financial risk protection, access to quality essential health-care services and access to safe, effective, quality, and affordable essential medicines and vaccines for all. Target 3.8 consists of two monitoring indicators: 3.8.1 for coverage of essential health services and 3.8.2 for catastrophic expenditure on health.

References

[1] Mueller KJ, Coburn AF, Lundblad JP, MacKinney AC, McBride TD, Watson SD. The high performance rural health care system of the future. Columbia, MO: Rural Policy Research Institute; 2011.

[2] Pong RW, Pitbaldo RJ. Don't take geography for granted! Some methodological issues in measuring geographic distribution of physicians. Can J Rural Med 2001;6:105.

[3] Institute of Medicine, Committee on Monitoring Access to Personal Health Care Services. Access to health care in America. Washington, DC: National Academy Press; 1993. https://www.ncbi.nlm.nih.gov/books/NBK235882/.

[4] Healthy People. Access to health services. Washington, DC: U.S. Department of Health and Human Services, Office of Disease Prevention and Health Promotion; 2020. http://www.healthypeople.gov/2020/topics-objectives/topic/Access-to-Health-Services. [Accessed 14 April 2016].

[5] AHRQ (Agency for Healthcare Research and Quality). 2009 National healthcare quality and disparities reports. Rockville, MD: AHRQ; 2010 [June 13, 2017] https://archive.ahrq.gov/research/findings/nhqrdr/nhqrdr09/qrdr09.html.

[6] AHRQ. 2016 National healthcare quality and disparities reports. Rockville, MD: AHRQ; 2017 [June 13, 2017] https://www.ahrq.gov/sites/default/files/wysiwyg/research/findings/nhqrdr/nhqdr16/2016qdr.pdf.

[7] Braveman P, Gottlieb L. The social determinants of health: it's time to consider the causes of the causes. Public Health Rep 2014;129(1_suppl2):19–31.

[8] Mikkonen J, Raphael D. Social determinants of health: the Canadian facts (PDF); 2010, ISBN:978-0-9683484-1-3. Archived (PDF) from the original on 2015-03-19. Retrieved2015-05-03.

[9] Commission on Social Determinants of Health. Closing the gap in a generation: health equity through action on the social determinants of health (PDF). World Health Organization; 2008, ISBN:978-92-4-156370-3. Archived (PDF) from the original on 2013-02-04.

[10] Wilkinson R, Marmot M, editors. The social determinants of health: the solid facts (PDF). 2nd ed. World Health Organization, Europe; 2003.

[11] Heiman HJ, Artiga S. Beyond health care: the role of social determinants in promoting health and health equity. Health 2015;20(10):1–10.

[12] Leong D, Roberts E. Social determinants of health and the Affordable Care Act. R I Med J 2013;96(7):20–2.
[13] Nicolette H. Social, economic and cultural environment and human health. In: Detels R, McEwen J, Beaglehole R, Tanaka H, editors. Oxford textbook of public health. 4th ed. New York: Oxford University Press, Inc.; 2004. p. 89–109.
[14] Wilkinson R, Marmot M, editors. Social determinants of health—the solid facts. 2nd ed. WHO-EURO; 2003.
[15] Piachaud D, Bennett F, Nazroo J, Popay J. Report of task group 9: social inclusion and social mobility. In: Task group submission to the marmot review; 2009. http://citeseerx.ist.psu.edu/viewdoc/download.
[16] Hayes A, Gray M, Edwards B. Social inclusion: origins, concepts and key themes: Australian Institute of Family Studies, http://apo.org.au/system/files/8799/apo-nid8799-90181.pdf; 2008.
[17] Mathieson J, Popay J, Enoch E, Escorel S, Hernandez M, Johnston H, et al. Social exclusion meaning, measurement and experience and links to health inequalities: a review of literature. WHO Social Exclusion Knowledge Network; 2008.
[18] Lenoir R. Les Exclus: Un Francais sur Dix. 1st ed. Paris: Editions du Seuil; 1974.
[19] The World Bank. Social exclusion and the EU's social inclusion agenda: paper prepared for the EU8 social inclusion study. The World Bank; 2007. http://siteresources.worldbank.org/INTECONEVAL/Resources/SocialExclusionReviewDraft.pdf.
[20] Peace R. Social exclusion: a concept in need of definition? Soc Policy J N Z 2001;16:17–36.
[21] Daly M. Social exclusion as concept and policy template in the European Union: Center for European Studies Working Paper Series #135: Minda de Gunzburg Center for European Studies Harvard University, https://ces.fas.harvard.edu/uploads/files/Working-Papers-Archives/CES_WP135.pdf; 2006.
[22] Ermann DA. Rural health care: the future of the hospital. Med Care Rev 1990;47(1):33–73.
[23] National Rural Health Association. Health care in frontier America: a time for change. Rockville, MD: Office of Rural Health Policy; 1994.
[24] Office of Technology Assessment. Health care in rural America. Washington, DC: U.S. Government Printing Office; 1990. U.S. Congress.
[25] Rosenblatt RA, Moscovice IS. Rural health care. New York: John Wiley & Sons; 1982.
[26] United Nations Children's Fund. United Nations system chief executives board for coordination, www.unsystem.org.

CHAPTER

6 Emergency disaster risk management for health

6.1 Introduction

Globally, people have been confronting with a wide and diverse range of risks linked to risks of health emergencies and disasters. Disaster is natural and also man-made. Infectious disease outbreaks, natural hazards, building collapses, chemical and radiation incidents, air pollution, flood, tsunami, earthquake, and jungle fire are some of the examples of devastating disasters overwhelmed the lifestyle with the tremendous socioeconomical crisis. The happenings of small-scale hazardous events with limited damage to the health system are common in different geographical locations, while great disasters have significant impact on public health, well-being, and development. These events resulted in devastating consequences, both in the acute phase and in the long term. In addition, the incident of climate change, population growth, unplanned urbanization, migration, and state fragility are catalyzing the frequency, severity, and impact of many types of emergencies throughout the world. It is the duty of government with public partnership to manage and find out innovative measures to ensure local, national, and global health security, to attain universal health care (UHC), and to develop the resilience of communities, countries, and overall health system. So, it is of utmost importance to develop sound risk management to safeguard the local community, national, regional [1, 2], and global strategies in the health system. It is important to adapt SDGs and bring UHC to strengthen the capacity of all countries, especially developing countries for early warning, risk minimization, and management of national and global health risk [3]; the Sendai Framework [4]; International Health Regulation (IHR) (2005) [5]; and the Paris Agreement [6].

The necessary for managing emergencies health-related risk in the disaster is critical for every nation. Every new threat in disaster challenges health and miserably affects life. Deaths, injuries, disabilities, psychological problems, and other health-related impacts in a disaster can be avoided or minimized by emergency health risk management measures. Emergency risk management for health is multisectoral and refers to the systematic analysis and management of health risk, posed by emergencies and disasters, through a combination of (i) hazard and vulnerability reduction to prevent and mitigate risk, (ii) preparedness, (iii) response, and (iv) recovery measures. Commonly, response to emergencies in a disaster situation focuses on primary health system. But, traditionally, it would not be possible to challenge the

Healthcare Strategies and Planning for Social Inclusion and Development. https://doi.org/10.1016/B978-0-323-90446-9.00006-X

broad spectrum of health risks caused by disaster. New and innovative ways and means have to be finding resilient health system for prevention and mitigation, and the development of community and nation capacities to provide a timely and effective response and recovery. Resilient health systems based on primary health care at the community level can minimize underlying vulnerability, protect health facilities and services, and scale up the response to meet the wide-ranging health needs in disasters.

6.2 What is a disaster?

In the Glossary of Humanitarian Terms, disaster is defined as: "A serious disruption of the functioning of a community or a society causing widespread human, material, economic or environmental losses which exceed the ability of the affected community or society to cope using its own resources" (International Strategy for Disaster Reduction, ISDR). Center for Research on the Epidemiology of Disaster defines a disaster as: "Situation or event, which overwhelms local capacity, necessitating a request to national or international level for external assistance." In 2002 WHO has defined "A disaster is an occurrence disrupting the normal conditions of existence and causing level of suffering that exceeds the capacity of adjustment of the affected community."

6.3 Types of disaster

Disaster is a short or long period of time that brings serious distraction in normal public life, social structure, environmental condition, socioeconomic condition, and various types of health problems [7, 8].

Here, we cover the public health disordered and management system by bringing resources to a normal condition suitable for tension-free life. Disasters occur in different forms with various time periods from few minutes to year(s) (Fig. 6.1A and B). On the basis of occurrence, disaster may be of natural origin or manmade (Fig. 6.2). On the basis of the intensity of damage, hurricanes and tropical storms are highly hazardous incidents but restricted to a specific geographical location. Some life-threatening contagious diseases also cause disaster and may last for more than a year like COVID-19 as a pandemic.

Disaster risk management is a practice for preventing the possibility of development of new risk, improving resilience to the effects of natural events, and contributing to sustainable development [9, 10].

The risk factors are variable in nature with the different population groups within a society or groups in specific geographic locations. The intensity of risk depends on causality, injuries, life-threatening diseases, disabilities, psychosocial problems, and other health impacts [11–13].

FIG. 6.1

(A) Type of natural disasters. (B) Man-made disasters.

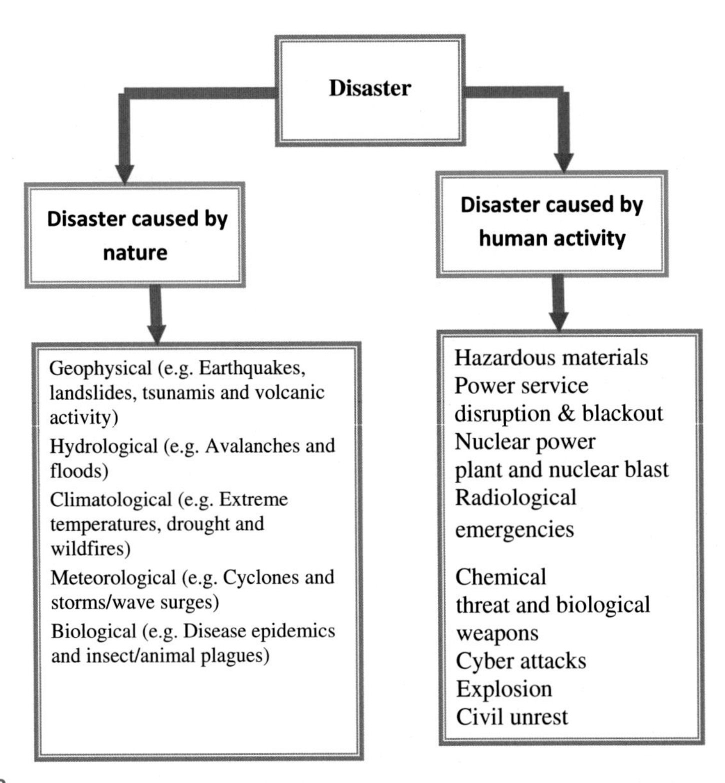

FIG. 6.2

Types of disasters.

To overcome such risks, a minimum time period depends on the potential of health workers, involvement of government, police framing, implementation of new rules and laws to integrate health into their risk management strategies.

This chapter is restricted to the public health approach to disaster risk management in order to reduce vulnerability by the implementation of prevention and mitigation measures. Besides traditional emergency care, it is also explained how to reduce the physical impact and to increase coping capacity and public awareness of the health sector and community.

6.4 Health consequences resulted from disasters

Globally, the most common hazardous events are transportation crashes, floods, cyclones/windstorms, jungle fires, industrial accidents, and earthquakes [14]. About 190 million people are directly victimized annually by emergencies due to natural and technological hazardous, with 77,000 deaths [14]. In addition, 172 million are affected by conflict [15]. Since 2012–2017, WHO recorded more than

1200 outbreaks. In 2018, a further 352 infectious disease events, including Middle East respiratory syndrome coronavirus (MERS-CoV) and Ebola virus (EVD), were tracked by WHO [16]. Mortality, morbidity, and disability resulted from these disasters caused severe disruptions of the health system and bring great disturbance in health service delivering through damage and destruction of the health facilities, interruption of health programs, loss of health staff, and overburdening of clinical services.

The financial strain developed during such disaster conditions is of great challenge for a country. Emergency caused by natural and technological hazards cost enormously high amount which brings turmoil in the financial condition of a country. About a decade back, the expected annual losses from pandemic risk through its effects on productivity, health system, and trade and travel have been calculated at about US$ 500 billion or 6% of global income per year [17]. It is estimated that premature deaths caused by air pollution cost the global economy about US$ 225 billion in lost labor income 2013 [18]. Commonly, most of the countries are likely to experience a large-scale emergency approximately every 5 years [19], and many are prone to the seasonal return of hazards such as monsoonal floods, cyclones, and disease outbreaks.

Both natural and manmade disasters have a direct impact on the health of the population at community and national levels. Disasters may increase the morbidity and mortality associated with contagious diseases through the impact on the healthcare system [20–22].

Most countries confront highly intensified disasters approximately within a gap of about 5 years. Many nations are prone to seasonal disasters such as cyclones, monsoonal floods, and disease outbreaks. But international attention focuses on high-consequence disasters resulted in mass-scale causality with tremendous damage in social habitation.

The most common disasters around the world are earthquakes, floods, and cyclones transportation crashes. It has been accounted that about 190 million people are directly victimized annually due to natural disasters, with about 77,000 deaths [23]. WHO report says more than 1200 outbreaks in 168 countries occurred from 2012 to 2017 [24]. In 2018, 352 infectious diseases events, including Middle East respiratory syndrome coronavirus (MERS-CoV) and Ebola virus diseases (EVD) were recorded [25]. Besides this, increasing morbidity, mortality, and disability emergencies also resulted due to severe disruption of the health system. It has been estimated that about US$ 300 billion annually cost occurred due to natural disasters [26]. Pandemic risk management costs about US$ 500 billion or 6% of global income per year.

6.4.1 What is health system?

A health system is referred to all organizations, institutions, resources, and people who are responsible for improving health services [27]. This is mainly targeted at improving the determinants of health. The health system is solely responsible for quality service, preventive measures for life-threatening diseases, promotive, curative, and rehabilitative interventions through a combination of public health action and the pyramid of health-care facilities that deliver personal health care by both state and nonstate actors. For effective function, a health system needs health workforces, funds, information network transport, communications, and overall guidance and direction to function.

6.4.2 Some important health hazardous

In the recent past, 2019, about 409 natural disasters were taken place worldwide. The Asian Pacific region was the most affected part causing due to the highest number of natural disasters. In 2018, Indonesia faced the most devastating earthquake and tsunami. Some of the important disasters faced by the people of the world are tabulated briefly in Table 6.1.

The year 2019–2020 faced the most horrifying times due to locus swarms, earthquakes, floods, landslides, and coronavirus. Followings are few such natural disasters described briefly:

(i) *Bushfires, Australia: 2019–2020*

In March 2020, about 18.6 million hectares of land and 5900 buildings (including 2779 homes) and killed at least 34 people due to bushfires. Nearly three billion terrestrial vertebrates alone were affected and some endangered species were believed to be driven to extinction. Economists estimated that the Australian bushfires may cost over A$ 103 billion in property damage and economic losses, making the bushfires Australia's costliest natural disaster to date. Nearly, 80% of Australians were affected either directly or indirectly by the bushfire (Fig. 6.3A).

(ii) *Flash flood, Indonesia*

In the early hours of 1st January, flash floods occurred throughout the Indonesian capital of Jakarta and caused the Ciliwung and Cisadane rivers to overflow. About 66 people have been killed and 400,000 displaced in the worst flooding in the area since 2007 (Fig. 6.3B).

(iii) *Covid-19, China all over the world*

In December 2019, the most deadly coronavirus was identified in Wuhan, China. China recorded the first death on January 11, 2020, and subsequently 2 months later, WHO declared the virus a pandemic on March 11. As April 2 came around, the world's number of cases surpassed a million. Currently, people are venturing out while taking precautions and trying to mitigate the horrifying situation slowly, with success. Countries such as India, Russia, and America have been desperately trying to develop vaccine(s) to resolve the pandemic situation at the earliest possible time (Fig. 6.3C).

(iv) *Volcano eruption, Philippines*

In January 12, 2020, Taal Volcano erupted resulting in the evacuation of about 8000 people at the beginning stage, but in due time it was closed by 3,00,000 people overall. About 39 death cases were noticed, but according to the Mania

Table 6.1 Important disasters faced by the people of world are tabulated briefly.

Nature of disaster and location/year	Damage caused
Haiti Earthquake (2010)	More than 220,000 people—2% or more of the population—were killed. One and a half million were displaced
Tōhoku Earthquake and Tsunami, Fukushima Daiichi Nuclear Disaster (2011)	As many as 20,000 people were killed A magnitude 9.0 earthquake off the coast of Japan triggered a tsunami wave that rose 133 ft at its highest and traveled as far as 6 miles inland
Hurricane Sandy (2012)	Causing widespread havoc through the Caribbean before crashing into the United States' eastern seaboard, taking large swathes of New Jersey and New York, including New York City, offline Over 100 people died in the United States alone, many from exposure or related conditions
Typhoon Haiyan (2013)	This Category 5 "super typhoon" crashed into the Philippines with wind speeds hovering near 200 miles per hour—at the time, the strongest cyclone ever. The storm killed approximately 7000 people and displaced more than 4 million
Nepal Earthquake (2015)	This magnitude 7.8 earthquake destroyed homes throughout much of the country and toppled tall buildings in Kathmandu, the capital. It's thought that the death toll—nearly 9000—could have been much higher
WEST AFRICA EBOLA OUTBREAK (2014–2016)	The deadliest Ebola outbreak in recorded history. The outbreak began in Guinea and quickly spread to Sierra Leone and Liberia- and striking heavily in urban centers. Ebola killed more than 11,000 peoples- approximately 40% of those who fell ill—over the course of 2 years
Hurricane Harvey (2017)	Harvey was a Category 4 storm with 130-mile-per-hour winds. Tens of thousands were displaced, critical access to health care was cut off and 88 people died
Hurricane Maria (2017)	When the devastating storm hit first Dominica (as a Category 5 hurricane) and then Puerto Rico (as a Category 4), it left devastation in its wake. As many of the 3000 deaths attributed to the storm
Cyclone Idai (2019)	The Category 3 storm crashed into southern Africa in March 2019. 1300 people were killed
Global Wildfires (2019)	Slash-and-burn agriculture caused massive, devastating wildfires in both the Amazon and Indonesia and destroying treasured forest and rainforest lands
COVID-19- CHINA-ALL OVER THE WORLD (2020)	In December 2019, the most deadly Coronavirus was identified in Wuhan, China. China recorded the first death on January 11, 2020, and subsequently 2 month later, WHO declared the virus to be world-wide pandemic on March 11. As April 2 came around, the world number of cases surpassed a million

Bulletin, people either perished because they refused to follow the evacuation order or decided to return to their homes, or died in the evacuation centers of heart attacks caused by anxiety (Fig. 6.3D).

(v) *Earthquakes, China-India-Iran-Philippines-Russia-Turkey*

The year 2019–2020, also faced a series of disastrous situations such as earthquakes all over the world. About 45 major earthquakes have been categorized over 6 degrees on the Richter scale. The most devastating earthquakes with the magnitude of 7 degree on the Richter took place in Jamaica, Russia, and Turkish city Izmir (Fig. 6.3E).

(vi) *Swarms of locusts, Asia-East Africa-India-Middle East-2020*

In 2020, locusts have swarmed in large numbers in dozens of countries, including Kenya, Ethiopia, Uganda, Somalia, Eritrea, India, Pakistan, Iran, Yemen, Oman, and Saudi Arabia. Desert Locusts are migratory pests that can eat as much food as around the amount 35,000 people can eat. They feed on crops and have the power to destroy crops within a short span of time. About 150 million of these pests can exist within a 1 km^2 area (Fig. 6.3F).

(vii) *Cyclone Amphan, Bangladesh-India-2020*

The Amphan cyclone is classified as one of the most powerful, deadly tropical cyclones to ever impact Bangladesh and India. It was categorized as a category 5 hurricane. It caused landfalls, heavy rain, gust windows, and lightning that left behind horrifying destruction and caused tremendous causality (Fig. 6.3G).

(viii) *Forest fires, Uttarakhand-2020*

In May 2020, Uttarakhand forest fire destroyed 71 ha of forest land and 2 people died. This could be one of the worst disasters faced by the state and India in general. The Kumaon region accounted for 21 forest fires, while the Garhwal region reported 16 forest fires. Uttarakhand is famous for its astonishing natural beauty is full of forests and wildlife. Backsides this, the forests of Uttarakhand are also home to numerous plant species, some of which are extremely rare (Fig. 6.3H).

(ix) *Assam floods, India-2020*

In May 2020, Assam faced a devastating flood from the Brahmaputra River in the Indian north-eastern state due to heavy rainfall. The flood affected over five million people, clamming the lives of 123 people, with an additional 26 deaths. About 5474 villages were affected and over 150,000 found refuge in relief camps (Fig. 6.3I).

(x) *Green snow, Antarctica-2020*

An increase in temperature has caused blooms of algae to multiply in the Antarctic. On the Antarctic Peninsula, the snow algae develop a thick layer of green patches. The snow algae thrive on temperatures just above freezing, which are increasingly common. The Antarctic Peninsula is one of the fastest-warming places on the Earth. The blooming of snow algae is acting as a bio-indicator showing the gradual raise in temperature around Antarctic Peninsula. Even a co rise in sea level has the potential to cause widespread flooding and soil erosion that could displace coastal communities permanently (Fig. 6.3J).

FIG. 6.3

(A–J) Major disasters during 2019–2020 all over the world.

6.5 Emergency strategies for disasters

The public of worldwide countries face a wide range of threats to health, due to infectious diseases outbreaks, unsafe food and water contamination, chemical and radiation contamination, natural and technological hazards, war and other social conflicts, and health consequences of climatic changes. In order to overcome, such challenges countries have to strengthen their capacities for health emergency and disaster risk management by adequate preventive majors, timely, and effective response and recovery. This could only be possible by the implementation of new need base policies that would be helpful for a resilient health system based on primary health care at the community level which can reduce underlying vulnerability, protect health facilities and services, and scale up the response to meet the urgency of wide-ranging health needs in disaster [17, 28–32].

Global health plays a significant role in the maintenance of global security as world economies become increasingly globalized. Due to extensive international travel and commerce, it is necessary to develop some international planning and frame rules in promoting a compressive, global, real-time infectious diseases surveillance system to avoid the possibility of transmission of contagious diseases threats.

Hence, health-care managers identify and evaluate risk in order to reduce injury to patients, staff members, and visitors within an organization. Health risk managers or paramedical trainers work force proactively and reactively to either prevent incident or minimize the damage of overall primary health problems caused in case of an incident of disaster.

But, a one-model-fits-all risk management system cannot be applicable to resolve solutions, resulted from a variety of disasters. So, administrators face challenges in risk assessment plan on the basis of circumstances generated in a disaster situation. Still, all countries should have proper multidisciplinary and multiple sectoral policies, strategies, and related programs to minimize health risks of emergencies, and disaster and their subsequent consequences.

Advocating for health crises resulted from disaster is not of one-man duty. It is a multisectoral partner's approach to mitigate the social, economic, and overall health damage caused by disaster by efficient risk management strategies. The multisectoral action plan should be focused on the generic elements of emergency risk management, including potential hazards, vulnerabilities of a population, and capacities, which can be applied across the various health domain. It is also necessary to identify the key points for consideration on a priority level for essential health domains.

A perfect strategic model is required to have a systematic approach for the real-time operating system for health risk mitigation caused by disasters at local, subnational, and national levels. Prior to launching new strategies, it is also necessary to examine and review existing plans and past experience which can be helpful in developing comprehensive strategies on a priority basis. Following are few such steps.

6.5.1 Role of health-care risk manager

Risk managers are well trained to tackle various issues in multiple settings. Following are some of the important responsibilities of health-care risk manager to minimize health risk-related problems.

Improve patient safety in hospital

In health-care risk management, first priority is to be given in the followings:

- use of latest confirmed technology easily available
- to convince patients about their treatment
- adaptation of protocols being approved by the regulatory agency to avoid medical errors
- follow proper handwashing procedure

The main focus is given on value-based health-care risk and time-line-based quality management system, and constant touch with patient safety issues. Proper collaboration of these two steps is the key to delivering safe, high-quality patient care with minimum risk. Followings are the few important steps to be carefully implemented by health-care managers:

Accreditation

Health risk managers are responsible for developing the framework to successfully lead the organization through the process of accreditation with various governing bodies in order to strengthen patient safety. This could only be possible by strengthening community confidence in the quality and safety of care, treatment, and services. Accreditation can be helpful in generating a competitive health-care environment and improve the ability to secure new business. In addition, accreditation can develop sustainability in health-risk management practice; reduce the liability insurance coverage; improve the efficiency of health-care workforce by getting carrier improvement training with updated information on primary health-care methods; provide customized intensive service information; enhance staff recruitment and development; and provide information to improve organizational structure.

In order to get a positive result from the health-care risk managers, the American hospital association offers different certification programs for health-care professionals. The American hospital association offers advocacy, research, services, and performance improvement support for health-care professionals and organizations. For example, they provide training for key health-care risk management issues, such as health-care reform, quality, and safety. But due to lack of financial budgetary provision and inadequate planning at the governance level, the developing countries fail to develop such practice at hospital and health institute levels.

Vendor risk management

Accreditation helps in real-time monitoring audits on third-party business associates, vendors, contractors, and employees, those are directly or indirectly associated with corporate sectors involved in the supply of health-care-related instruments for body

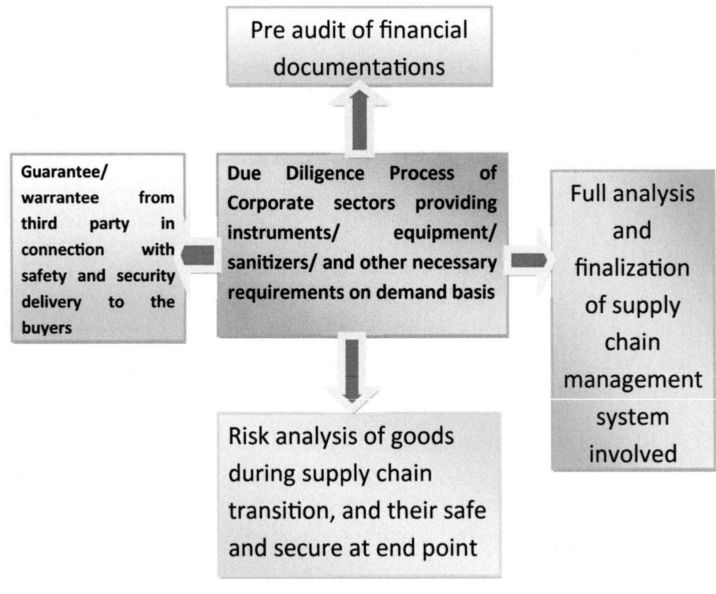

FIG. 6.4

Corporate due diligence process for quality health care-related goods.

check-ups, sanitizers, ventilators, etc. The practice of vendor risk management is necessary and helpful in maintaining corporate reputation. In addition, risk audit of third-party vendors is vital to protect from harm, and uphold corporate due diligence (Fig. 6.4).

This is because risk managers are solely responsible for regulating the protocol and policies related to the hiring practices involving external vendors, business associates, and other stockholders such as human resources, supply chain, legal, and material management.

Financial risk

Commonly, health-care risk managers are not associated with negotiations or administration of contracts, and as a result, the administrators are not well convincing with the faults committed by the vendors or contractors while dealing with health-care contracts. Revive of the contract without proper assessment may lead to tremendous health risk problems. So, it is necessary for a health risk manager to understand the action plan on how contracts are managed.

Data security

Health-care data security is an important issue of health insurance policy framing and making it update. Well and updated data on health risk survey systems should be available to health risk managers which would be helpful to keep vigilance on any mistakes or irregularities while implementing health-care risk policies matter.

Under the prevailing condition of the COVID-19 infectious diseases during the pandemic, the main challenge in health-care risk is the cyber risk. The World Health Organization has noticed a dramatic increase in the number of cyberattacks. For example, in Italy COVID-19 emergency has tremendously affected cybersecurity; from January to April 2020, the total of attacks, accidents, and violations of privacy to the detriment of companies and individuals has doubled. But, due to the lack of proper attention of health-care managers on the topic the cybersecurity and management are in alarm condition.

6.6 Why risk management

Due to changes in climate and enormous loading of pollutants on the surrounding, natural, chemicals, biological, and societal hazards bring tremendous adverse effects on public health. Examples of these hazards are given as follows: natural calamity, earthquake, landslides, tsunami, cyclone, flood, and draft. This misshapenness causes ill health directly or through the disruption of the health system, facilities, and services, leaving many without access to health care in times of emergency. In addition, they affect public health facilities such as water supplies and safe shelter.

So, sustainable development goals (SDGs) have kept special provisions under SDG3 to "ensure healthy lives and promote well-being for all at ages." The SDG declaration emphasizes that to achieve the overall health goal, "we must achieve universal health coverage (UHC) and access to quality health care, No one must not leave behind." SDG 3 aspires to ensure health and well-being for all, including a bold commitment to end the epidemics of AIDS, tuberculosis, malaria, and other communicable diseases by 2030.

6.6.1 Integrate risk management of health

In order to strengthen the health system, it is necessary to adapt the IHR (2005), and planning multihazardous disaster risk management strategies are supposed to be the milestone of progress to mitigate the health risk problem associated with disaster situation. Generally, random approaches to resolve the health risk problem in a disaster situation may not be successful due to the lack of proper coordination between different stakeholders. So, to bring success in health risk management, it is necessary to have an innovative integrated approach by bringing perfect mutual understanding among different health workforces. There is a need to frame a comprehensive contemporary approach and practice through the conceptual framework or paradigm of "health emergency and disaster risk management."

Integrated healthcare (integrated health or coordinated care) is an approach characterized by highly efficient teamwork and communication among health professionals (Fig. 6.5). An integrated approach to health risk has become a worldwide trend in health-care reforms and new organizational arrangements focusing on more coordinated and integrated forms of care provision [33–35].

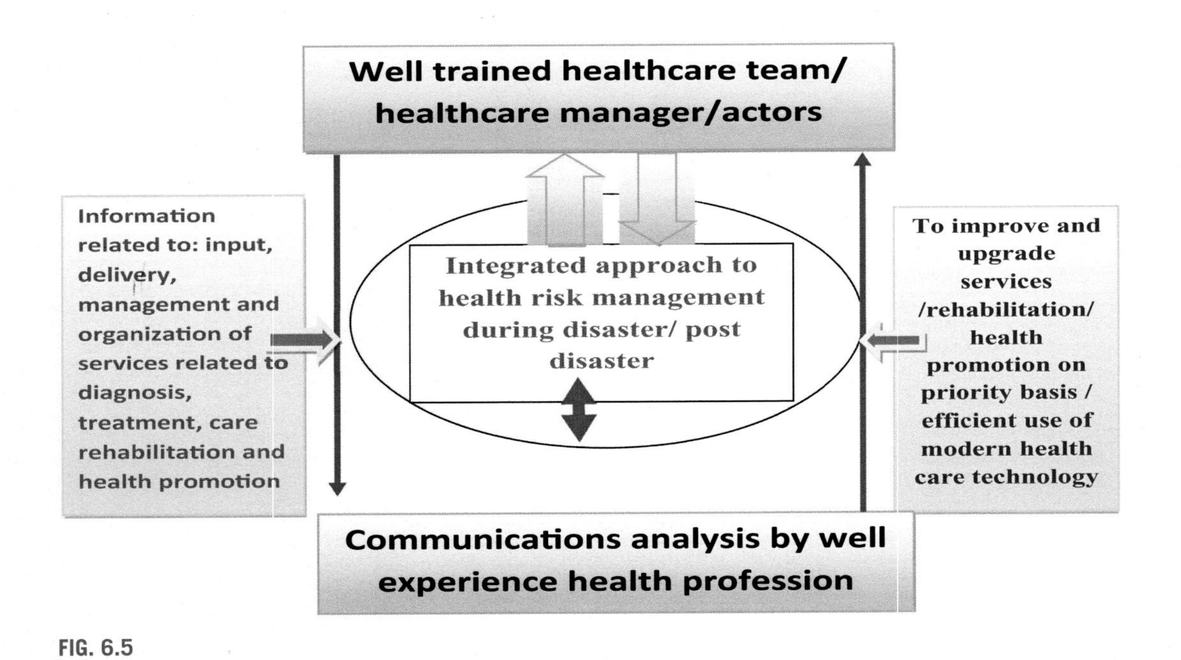

FIG. 6.5

Schematic representation of integrated risk management of health.

World Health Organization explains integrated health risk care as: "Integrated care is a concept bringing together input, delivery, management and organization of services related to diagnosis, treatment, care rehabilitation and health promotion. Integration is a means to improve services in relation to access, quality, user satisfaction and efficiency." The process of integration may be in the horizontal form (linking similar levels of care like multiple professional teams) or in the vertical form (linking different levels of care like primary, secondary, and tertiary care).

Proper risk management strategies are the key to minimize the consequences of emergencies in natural disasters. Risk is defined as "The combination of the probability of an event and its negative consequences." In more detail, risk can be explained as "The potential loss of life, injury, or destroyed or damaged assets which could occur to a system, society or a community in a specific period of time, determined probabilistically as a function of hazard, exposure, vulnerability and capacity." It is not possible to completely resolve the hazard-related risk, but can minimize the damages related to normal health.

In response to a wide range of disasters such as natural, biological, technical, and community hazards, a range of risk management practices are implemented. Both the primary prevention and recovery emergency recovery awareness are also taken care in order to minimize the health and other consequences of emergencies at the community and state level. While implementing such prevention and recovery majors, it is also necessary to take care of cost-effective strategies in comprehensive risk management form.

Integrate risk management of health during natural or manmade disasters is supposed to be a set of practices and processes supported by a team of well trainee health-care managers who can handle efficiently the latest health-care equipment, instruments, sensitizers, ventilators on the emergency need base during disaster or post-disaster situation prevailing in a specific geographical location(s). So, it is necessary to frame a health risk management strategy in a hospital in order to attend emergency situation arises during the disaster (Fig. 6.6).

Following are few important aspects to be observed by health risk managers.

(i) *Strategy*

Implementation of a well-framed blueprint on structures, roles, and responsibilities of governments and nonvoluntary organization (NGO) in handling emergency risk management for health on the basis of the type of disaster.

(ii) *Assessments*

Identification, evaluation, and prioritization of risk for early warning and surveillance; technical guide dance; information management; and public communication.

(iii) *Responses*

Identification and recognition of a wide range of health-care services and related measures to mitigate the health risk.

(iv) *Health infrastructure and logistics*

Development or arrangement of temporary infrastructure, and supply chain management for the distribution of health-care-related goods, in a safe and timely manner to support health-care risk management is the most challenging aspects during a pandemic or disease outbreak.

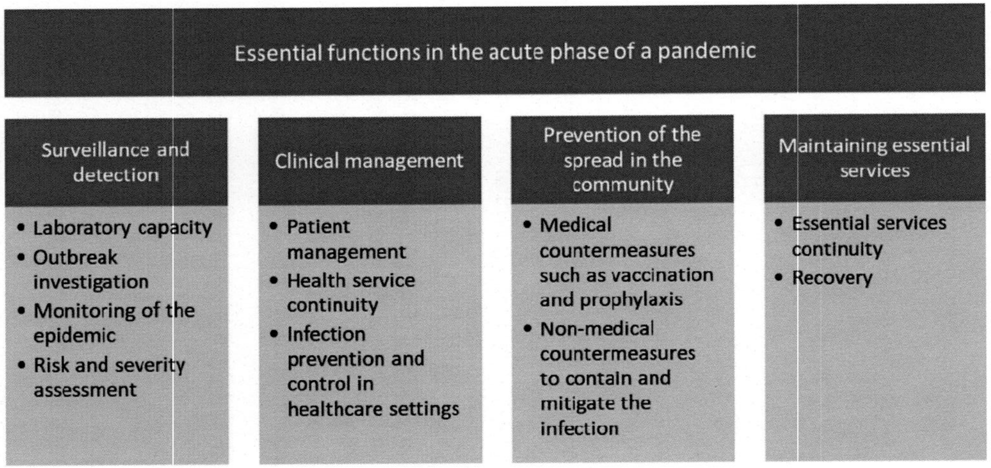

Essential functions in the acute phase of a pandemic

Surveillance and detection
- Laboratory capacity
- Outbreak investigation
- Monitoring of the epidemic
- Risk and severity assessment

Clinical management
- Patient management
- Health service continuity
- Infection prevention and control in healthcare settings

Prevention of the spread in the community
- Medical countermeasures such as vaccination and prophylaxis
- Non-medical countermeasures to contain and mitigate the infection

Maintaining essential services
- Essential services continuity
- Recovery

FIG. 6.6

Flattening the COVID-19 peak: Containment and mitigation policies.

Prepared by OECD based on World Health Organization (2005 [15]) "WHO checklist for influenza pandemic preparedness planning."

FIG. 6.7

A photographic view of the World's Sports Stadium converted into a temporary hospital (China).

For example, in the case of large-scale emergencies such as spread of contagious diseases like COVID-19 for a prolonged period is most challenging. Immediately after the large-scale outbreak, Wuhan adapted three methods to manage and reduce the alarming conditions resulted from COVID-19. The government adopted three methods to admit patients for treatment: started building temporary hospitals at a distance from the city territory; designated hospitals; and fenced shelter hospitals (by converting large-scale public venues such as stadium and exhibition centers) (Fig. 6.7). These steps were taken mainly for isolation, treatment, and disease monitoring of patients with mild symptoms.

(v) *Health and related services*

Identification and operation of processes which can systematically track governance objectives, risk ownership or accountability, compliance with policies and decisions framed by the government, related to the effectiveness of risk mitigation and control.

This overall approach should be undertaken by the government under all types of disasters (e.g., biological, geological, chemical, hydrometeorological, societal) regardless of the causes that arise during disaster and post-disaster period. In addition, it is necessary for countries to have thorough review for strengthening at all levels of care, and to build capacities for reducing the risk of future emergencies. It is also a desire to develop and establish a common language which can be helpful for those health-care working forces involve in mitigating health risk issues emerging due to disaster conditions.

6.7 Components and function of EDRM

The components of EDRM have a wide range of functions in health and other related sectors which have a significant role in managing the health risks of emergencies and disasters. From functional aspects, all risks are collectively managed and effectively coordinated for successful Health EDRM. The Health EDRM functions are categorized into the following components, derived from a number of sources including the adaptation of the health system building blocks, multisectoral emergency and disaster management, and the IHR (2005) including epidemic awareness and response.

6.7.1 Legislation strategies

The member of states should incorporate national health policies, strategies, and plans, supported by appropriate legislation with adequate budgetary provision to resolve the variety of health risks problems resulted from the disaster. In addition, framing of national policy should be made keeping in view on all public, private, and civil society stakeholders, across the components of all-hazards Health EDRM. This is also applicable to multisectoral EDRM policies and legislation that should refer to the protection of people's health and the minimization of health consequences as specific aims and outcomes.

6.7.2 Planning and coordination

Proper plans are necessary to execute Health EDRM. Additionally, special emphasis has to be given to the agendas of IHR (2005) and the Sendai Framework. It is also important to assess risk and capacity analysis, exercise, and reviews, especially those conducted for national multisectoral all-hazards disaster risk management under the IHR Monitoring and Evaluation Framework. The health and multisectoral plan should be integrated with health security, national disaster risk reduction plans, plans for preparedness, response and recovery, and incident management system.

While developing such an integrated planning system, intense care is to be taken to link with jurisdictions—local, subnational, and national. Time and again, the plans for emergency preparedness and response need to be regularly tested and reviewed. During EDRM, it is also important to take care of business continuity plans to ensure that vital functions and services continue throughout the emergency [36, 37]. Health EDRM coordination mechanisms and/or dedicated units should be established to ensure appropriate coordination across the health sector and with other sectors at each level.

6.7.3 Man powers

Well-trained health workforce must be developed to manage Health EDRM operational strategies effectively at national, subnational, and local levels. The health workforces are to be well informed about the significance of EDRM for community well-being. Skilled human resources are the core workers to handle a wide spectrum of health risk management such as emergency planning, incident management, epidemiology, laboratory diagnostics, information management, risk, and needs assessments, logistics, risk communication, and health-care delivery.

6.7.4 Budgetary provision

Generally, in a developing country, budgetary resources often fail to fund measures to proactively reduce the risks of disaster. Still, the government tries to divert large sums responding to emergencies, particularly high-impact disasters such as West African Ebola outbreak of 2014–2016 or from the cluster of storms that hit the Gulf of Mexico during a few deadly weeks in 2017 or the recent COVID-19 pandemic which brought turmoil in world economy.

In order to meet the emergency of Health EDRM, it is of utmost importance that the government should have adequate special budgetary provision to implement programs and activities at the state level, effectively. The budgetary provision should be flexible to meet the expenses for health EDRM on the need base.

6.7.5 Information and communication

Information and communication technology plays a significant role in disaster prevention, mitigation response, and recovery. Timely predictable and effective information is much needed by government agencies and other humanitarian actors involved in rescue operations and decision-making processes. It is important that information collection, analysis, and dissemination are harmonized across relevant sectors, and mechanisms put in place to ensure that "the right information gets to the right people at the right time."

6.7.6 Information and knowledge management

It is necessary to strengthen information and knowledge management capacities to support risk assessments, disease surveillance and other early warning systems, and public communications. It is also important that information collection, analysis, and dissemination are harmonized across relevant sectors; research supports the evolution of evidence, knowledge and practice, and the development of new drugs, vaccines, and innovative risk management measures.

6.7.7 Risk communications

Currently, risk communication has a significant role in Health-Emergency Disaster Risk Management (Health-EDRM). Risk communication target is to empower communities to invest and emphasize their disaster health risk reduction effects, thereby strengthening health systems and supporting community health resilience building. Risk communication is a critical function of Health EDRM, especially when relating to other sectors, government authorities, the media, and the public. Real-time access and exchange of information are vital so that everyone at risk is able to make informed decisions and take action to prevent, mitigate, and respond to potential emergencies. The stakeholders should coordinate properly on the basis of public information in order to avoid conflicting information being disseminated.

6.7.8 Health infrastructure and logistics

Many basic services such as water, sanitation, and energy (management of electricity), which are directly responsible for health system management should be adequately available and continue to function before, during, and after disaster situations. Supporting logistics will include stockpiling, and prepositioning of medicines and supplies, effective supply chains, and reliable transportation and telecommunication system [38, 39]. The only means to protect life in health-EDRM is to provide quality health facilities and related infrastructure for the safe and security of public life.

6.7.9 Health and related services

In order to develop quality health services, it is necessary to pay primary attention to prehospital and facility-based clinical services in the event of an emergency with health consequences. During health EDRM, there should be adequate service delivery to meet increased health needs by increasing bed capacity, establishing temporary facilities, mobile clinics, and vaccination campaigns.

6.7.10 Community capacities for health EDRM

Participation of communities in risk assessments to identify local hazards and vulnerabilities can identify actions to reduce health risks prior to an emergency occurs. Many lives can be saved in the first hours after an emergency through effective local response before external help arrives. The local population will also play the lead role in recovery and reconstruction efforts. Community capacities and activities including primary health care and the role of local health workers, civil society, and the private sector are therefore central to effective Health EDRM. Civil society can be well trained to tackle issues such as community-level surveillance, household preparedness, local stockpiling, first aid training, and emergency response. In this connection, ministries and the private sector may be responsible for managing critical infrastructure (e.g., water supply, electricity, transport, and telecommunications) and contribute to civic activities.

6.7.11 Monitoring and evaluation

It is primarily important to have control over monitor progress toward meeting health EDRM objectives and core capacities should be integrated into existing health sector monitoring systems. It is of utmost importance to develop standardized relevant indicators include the Sendai Framework Monitor for targets and indicators, IHR Monitoring and Evaluation Framework, WHO global survey on country capacities for Health EDRM, and WHO regional monitoring and evaluation mechanisms.

6.8 Basic strategies for health risk management

6.8.1 Target

The basic target for health risk management is to organize a well-trained qualified health-care risk manager to implement and operate the blueprint plans framed and approved by the government. While operating the health-care risk management plan, the health-care manager should have priority basis approaches such as finance, safety, and patient care. So, the vision of the World Health organization is "the highest possible standard of health and well-being for all people at risk of emergencies, and stronger community and country resilience, health security, universal health coverage and sustainable development."

6.8.2 Duties of health-care risk managers

Health-care managers identify and evaluate risks in order to mitigate or prevent various emergency health issues caused due to disasters. The duties and responsibilities given to health-care risk managers are ultimately determined by the government or by the specific organization, and be guided by the following core principles and approaches:

Risk-based approach

Risk-based health services are the most critical in order to find out proper rehabilitation of the people victimized due to disaster. In such cases, continuous vigilance and patients care to require on the most priority basis around the clock all 365 days. The health risk that resulted from a disaster depends on the exposure to hazards, their vulnerabilities to those hazards, the intensity of the hazardous condition. So, it is necessary to understand and how to undertake the emergency operation in the hospital to manage health risks caused by the disaster.

Compressive emergency management

By means of interrelated and comprehensive approaches, the health risk operation can be mitigated the intensification of hazardous conditions at community and country levels. The compressive approach means a sequential closely interrelated prevention/mitigation emergency awareness, responses, and recovery measures in order to reduce the likelihood and severity of health risk emergencies.

All-hazards approach

All-hazards health risk assessment refers to an integrated approach to assist facilities in identifying the greatest threats and vulnerabilities in the full spectrum of emergencies or disasters, including internal emergencies and manmade.

Often, different types of hazards are linked with similar health risks, and the overall operating system to mitigate the health risk problems. It is not advisable to plan separate, stand-alone capacities, or response mechanisms for individual hazards. So,

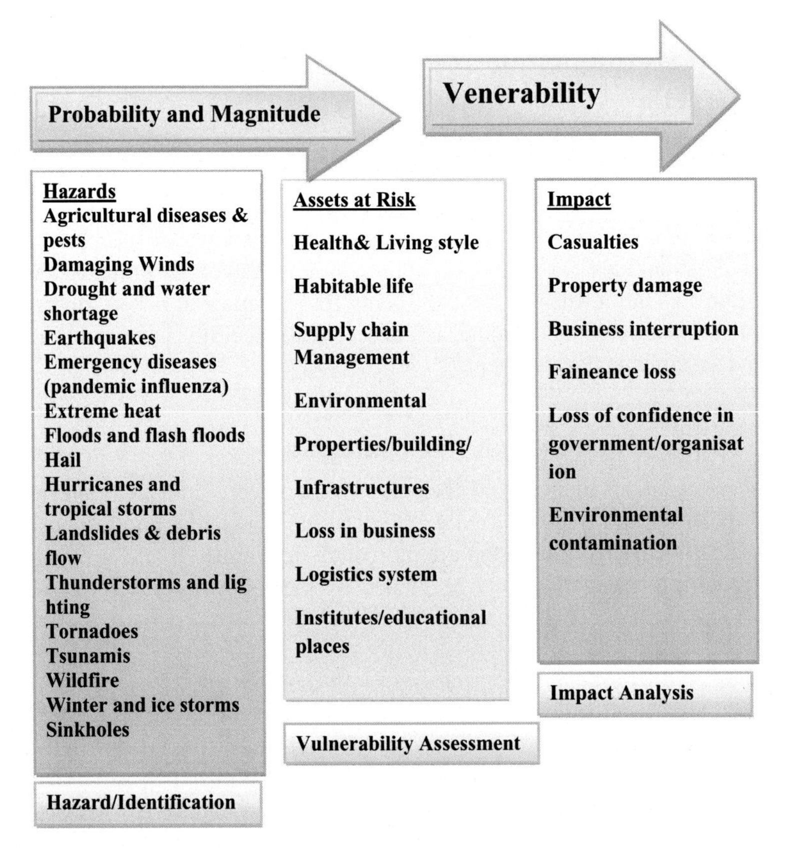

FIG. 6.8

All-hazards approach to physical security planning.

it is necessary to frame common health-related emergency disaster risk management policies under specific and limited budgetary provisions (Fig. 6.8).

People- and community centered

Commonly, the health livelihood in the community at the village level is more victimized than urban areas during the disaster period. The village and rural communities are the first responders to an emergency caused by hazardous conditions resulted from the disaster. It is desirable that an inclusive approach, based on accessible and nondiscriminatory participation, is necessary to mitigate health risk-related hazardous. Mainly, women, children, people with disabilities, older persons, migrants refugees and displaced persons, people with chronic diseases are to be taken care of on a priority basis. Emergency disaster risk management should be integrated gender, age, disability, and cultural perspectives, in which village youth and women should be given the opportunity to drive the health-care risk management operation system.

Multisectoral collaboration

Multisectoral collaboration refers to the systematic investigation, analysis, and management of health risks resulted due to disaster by undertaking the issues in a combination of: (i) minimization of vulnerability, (ii) awareness (iii) response, and (iv) recovery measures. A strong intersectoral approach is based on collaboration, communication, and coordination across public health, animal health, and other related sectors and disciplines to efficiently manage health risk at the human-animal-environment interface with the target to achieve maximum health risk recovery at the community and country level. In a multisectoral approach the overall responsibility in the disaster period is undertaken with strong mutual understanding between the health sector (for managing the risk of infectious diseases), meteorological services (for timely prediction and warning related to weather conditions), civil protection (for an emergency response to disaster hazardous), food security, and supply chain management for health-related goods and medicines. Besides these, emergency disaster risk management activities need the help of other sectors, e.g., infrastructure management, water and sanitation for human needs and functioning of health facilities, transportation, logistics, emergency services, and food security.

Ethical consideration

Ethical issues involved in disaster risk management are critical aspects for health-care manager or actors to assess themself about right versus wrong conduct and what constitutes a good or bad life. The overall success in health risk management mainly depends on how health workers view and treat victims and promote justice in society and institutions. Ethics includes justifying irrespective of poor and rich or influenced personalities while discharging the responsibility of health workers and actors in order to mitigate health risks problems resulted in natural or manmade disasters. While serving mankind in order to mitigate health risk problems the health-care workforces should give justification for values and beliefs that are deeply rooted in the cultural, political, religious, and philosophical way of life at the community level.

6.9 Disaster management cycle to mitigate health risks

Disaster health management is the discipline dealing with minimizing health-related problems to give relief. Mainly, the disaster management cycle is to support rebuilding society suffering from health risk problems in a disaster situation (Fig. 6.9).

Effective emergency health risk management depends on the integrated action plan of government or internationally repute NGO, based on the intensity of health hazards caused by the outbreak of disaster [40–42].

Followings are the few important majors taken to give relief to the people from the health risks on phase wise.

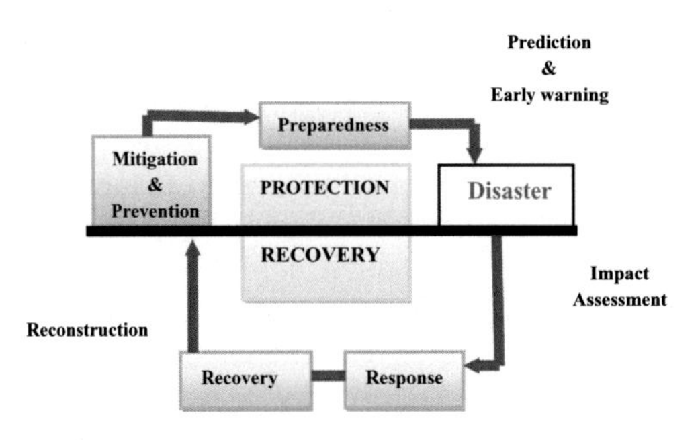

FIG. 6.9

Pictorial representation of disaster management cycle to mitigate health risks.

6.9.1 Mitigation

The events caused by a disaster have both positive and negative effects on various issues related to health. The impact and interventions to reduce the negative effect are known as risk mitigation. So, disaster health risk mitigation refers to minimize the health hazardous or to eradicate the health-related issues completely. The mitigation phase differs from the other phases because it focuses on long-term measures for reducing or eliminating risk at personal, family, and community levels. The overall operational process for health risk mitigation can be categorized under the following heads: (i) assessment (type of health risk and intensity), (ii) planning (priority-based action plan), (iii) response to an event (occurrence of event and its possible continuity), and (iv) evaluation of the response (positive/negative aspects of recovery from health risk).

The action plans taken for mitigating the health risk of a disease outbreak are

- education of all employees on personal protective equipment such as gloves, masks, gowns, and respiratory masks
- strict handwashing monitoring in all units
- education related to isolation procedures
- monitoring patient and staff vaccinations
- education for all patients in various forms of multimedia

6.9.2 Preparedness

The preparedness phase refers to the preparation and development of blueprint of health risk management by the health-care managers for the disaster strike under the following action plan:

- communication plans with commonly used terminology and sequential commands to carry out risk management operations timely and efficiently
- coordination of multiple agencies engage in primary health and risk management

- frequent training for emergency services
- arrangement of temporary health-care center to prevent further contamination
- arrangement of the proper supply chain for delivering health-care goods and medicines, timely

An efficient preparedness measure is to develop special emergency operations center (EOC) with the help of local or regional doctors to manage emergency health risk problems. In addition, it is also necessary to train local volunteers and deploy them to assist health-care managers or actors. The local volunteers can be helpful in the construction of shelters, arrangement of warning devices, backup life-line services (e.g., power, water, sewage), and rehearsing an evacuation plan.

6.9.3 Response

The response phase refers to the effectively mobilization of the necessary emergency services and first responders in the disaster area. The first core emergency services such as fire-fighters, police, and ambulance crews are to be managed with the support of a number of secondary emergency services, such as specialist rescue teams.

6.9.4 Recovery

The recovery phase is referred as how much control acquired over the disaster-based health causality and restoring their lives and infrastructure that support them. But it is hardly to assess at what junction immediate relief changes into recovery and long-term sustainable development. During the recovery phase, there would be many opportunities to enhance prevention and increase preparedness to reduce the intensity of vulnerability. Recovery activities continue until all systems to normal or better. The short- or long-term recovery measures include: returning vital life-support systems to minimum operating standard; temporary shelter; an adequate supply of public information; health and safety education; reconstruction; counseling programs; and economic impact studies.

6.10 Stakeholders involve in disaster management

A wide range of sectors and stakeholders are involved in disaster risk reduction policies and strategies to prevent the strengthening of resilience and reduction of disaster losses. The roles of some key stakeholders are outlined below.

6.10.1 Government and society

The role of local government and NGOs have critical in dealing with risk reduction in a disaster situation to build resilient communities which can well survive in post-disaster conditions. Because the local government has first-hand knowledge of the community's social, economic, infrastructure, and environmental needs may be helpful to provide support in a disaster.

The Hyogo Framework for Action 2005–2015 (Kobe, Japan) passed the resolution that both the community and local authorities should be empowered to manage and minimize disaster risk by having access to the necessary information, resources, and authority to implement actions. In this connection, it is necessary to bring reformation and improvement of poor urban governance, informal settlement on unsafe land, declining ecosystem, and vulnerable rural livelihood. In order to boost such activity, the United National Office for Disaster Risk Reduction (UNDRR) was created in December 1999 to ensure the implementation of the International Strategy for Disaster Reduction. UNDRR (formerly UNISDR) also supported the implementation and review of the Sendai Framework for Disaster Risk Reduction adopted by the Third UN World Conference on Disaster Risk Reduction on March 18, 2015, in Sendai, Japan. The main target of achievement of Sendai meet is to minimize in reduction of disaster risk and losses in lives, livelihoods and health and the economic, physical, social, cultural, and environmental assets of persons, businesses, communities, and countries over the next 15 years. It is necessary to have the well link of the health sector with other sectors in order to strengthen and managing health risks of emergencies at local, national, and international levels [43, 44].

6.10.2 Role of health sector

The Ministry of Health at the national and subnational level has the main responsibility in emergency disaster risk management measures related to the outbreak. It is the primary responsibility of the ministry of health to address and coordinate with the national and international disaster risk organization to frame health-care task force with the help of local NGOs to workout effectively to minimize health risk by following the blueprint framed by WHO related to emergency disaster risk management strategies.

Reducing the health risks and consequences of emergencies is vital to local, national, and global health security and to develop the resilience of communities, countries, and health systems.

6.10.3 National disaster agency

At the country level, disaster risk management is taken by their respective country-based organization to immediately undertake large-scale emergencies and disasters caused by natural or manmade disasters. The National disaster authorities should ensure fully with health-care workforces to meet emergency risks caused by a specific hazardous incident. In case of an alarming uncontrolled situation, they should ask the help of international disaster management organization as per the understanding of Sendai Framework and SDGs campaign for health risk reformation and sustainable normal livelihood.

6.10.4 World Health Organization

World Health Organization (WHO) in association with national and international health partners supports and strengthens the health risk campaign and action in disaster affected locations at the state and country levels. So, WHO developed the "health emergency and disaster risk management framework (EDRM)" to extend help to ministries of health and other stakeholders of different countries to help in implementing the following:

- planning and coordination
- human and financial resources
- information and knowledge management
- risk communication
- community capacities for health EDRM
- monitoring and evaluation

WHO provides support for the international frameworks such as the Sendai Frame for Disaster Risk Reduction 2015–2030 and other UN system policies plans.

WHO also safeguards the implementation of some international campaigns such as the Sustainable Development Goals (SDGs), including the pathway to universal health coverage. In this connection, WHO provides technical support on building capacity leadership for articulating policy options such as technical guidance, norms, and standards, designing research and development, and monitoring health trends in terms of risks, health effects, and country capacities.

In order to the successful implementation of health risk management, WHO collaborates with many stakeholders in emergency health risk management at national, subnational, and local levels, international organizations, multilateral and bilateral agencies, civil society, and nongovernment organizations, research, and educational institutions. Followings are few important international summits in which WHO has also played a key role in the implementation of disaster risk management at national and international levels (Table 6.2).

The WHO is associated with many international agencies, intergovernmental organizations, International Red Cross and Red Crescent Movement, NGOs, and the private sector in the disaster risk eradication movement. For example, the Capacity for Disaster Reduction Initiative (CADRI) is a global partnership consisting of 16 UN and non-UN organizations involve in strengthening the capacities to minimize, and even prevent risk resulted from disasters.

6.10.5 The International Health Regulation

The International Health Regulation (IHR) was first adopted by the World Health Assembly in 1969, and last revised in 2005. It is a legally binding instrument of international law with the target of international collaboration "to prevent, protect against, control, and provide a public health response to the international spread of disease in ways that are commensurate with and restricted to public health risks and that avoid unnecessary interference with international traffic and trade" [45–47].

The IHR is the only international legal treaty with the responsibility of empowering the World Health Organization (WHO) to act as the main global surveillance system [48, 49].

Table 6.2 List of some international disaster risk management summits.

Year	Location	Conference
2006	Davos, Switzerland	1st International Disaster Reduction Conference—IDRC Davos 2006
2007	Harbin, China	1st regional International Disaster Reduction Conference—IDRC Harbin 2007
2008	Davos, Switzerland	2nd International Disaster Reduction Conference IDRC Davos 2008 "Public-Private Partnership—key for integral risk management and climate change mitigation and adaptation"
2009	Chengdu, China	2nd regional International Disaster Reduction Conference IDRC Chengdu 2007 "Wengchuan Earthquake, the path forward"
2010	Davos, Switzerland	3rd International Disaster Reduction Conference IDRC Davos 2010 "Risk, Disasters, Crisis and Global Change—From Threats to Sustainable Opportunities"
2012	Davos, Switzerland	4th International Disaster Reduction Conference IDRC Davos 2012 "Integrative Risk Management in a Changning World—Pathways to a Resilient Society"
2014	Davos, Switzerland	5th International Disaster Reduction Conference IDRC Davos 2014 "Integrative Risk Management—The role of science, technology and practice"
2015	Sendai, Japan	World Conference for Disaster Risk Reduction
2015	Japan	Sendai Framework for Disaster Risk Reduction 2015–2030
2017	Cancun, Mexico	Health at the 5th Session of the Global Platform for Disaster Risk Reduction
2019	Geneva	Health at the 6th Session of the Global Platform for Disaster Risk Reduction

The IHR[a] (2005) contains a wide range of innovations, including:

a. Scope of IHR is not limited to any specific disease or manner of transmission bur covering "illness or medical condition, irrespective of origin or source that presents or could present significant harm to humans";
b. It is the prerogative for a member of state to develop certain minimum core public health capacities;
c. Obligations on State Parties to notify WHO of events that may constitute a public health emergency of international concern according to defined criteria;
d. Provisions authorizing WHO to take into consideration unofficial reports of public health events and to obtain verification from States Parties concerning such events;
e. Procedures for the determination by the Director-General of a "public health emergency of international concern" and issuance of corresponding temporary

[a] Source: quoted from available on the WHO website (www.who.int).

recommendations, after taking into account the views of an Emergency Committee;

f. Protection of the human right of persons and travelers; and

g. The establishment of National IHR Focal Points and WHO IHR Contact Points for urgent communications between States Parties and WHO.

6.10.6 International community

The WHO is associated with many international agencies, intergovernmental organizations, International Red Cross and Red Crescent Movement, NGOs, and the private sector in the disaster risk eradication movement. For example, the Capacity for Disaster Reduction Initiative (CADRI) is a global partnership consisting of 16 UN and non-UN organizations involved in strengthening the capacities to minimize and even prevent risk.

6.11 COVID-19: Strategies update and management

As of April 13, 2020, more than 1.7 million people were under the severe grip of COVID-19, and out of which 85,000 people lost their life [50].

The pandemic COVID-19 overwhelmed the world health system and resulted in tremendous damage to social and economic structure. Meanwhile, world scientific community meticulously analyzed the data related to COVID-19 pandemic and are still busy further analyzing the ongoing scenario of corona infection, and health measures are taken to understand clearly how it is spreading, the severity of diseases it causes, how to tackle it, and how to eradicate. This pandemic is much more than a health crisis. It requires the whole of government and whole-of-society response.

COVID-19 is a new disease, distinct from other diseases caused by coronaviruses such as severe acute respiratory syndrome (SARS) and middle east respiratory syndrome (MERS). The virus is highly contagious, and spreads exponentially like jungle fire. The COVID-19 pandemic has taken a huge toll on individuals, families, communities, and societies across the world. Daily lives have profoundly changed; economies have fallen into recession, and many of the traditional social, economic, and public health have lost their respective ability to have control over the COVID-19 pandemic. The public has lost their confidence in the health safety networks of both private and government sectors.

The overall mortality rate due to the COVID-19 pandemic varies remarkably, depending on the geographic location of the country, the status of primary health centers, communication system, and financial status of state government.

Within a short span of time, the outbreak of COVID-19 resulted in the pandemic, mainly due to

(i) The spread of disease like jungle fire to each nook and corner of the world and has overwhelmed the health system, including social structure and function of community and social life (Fig. 6.10).

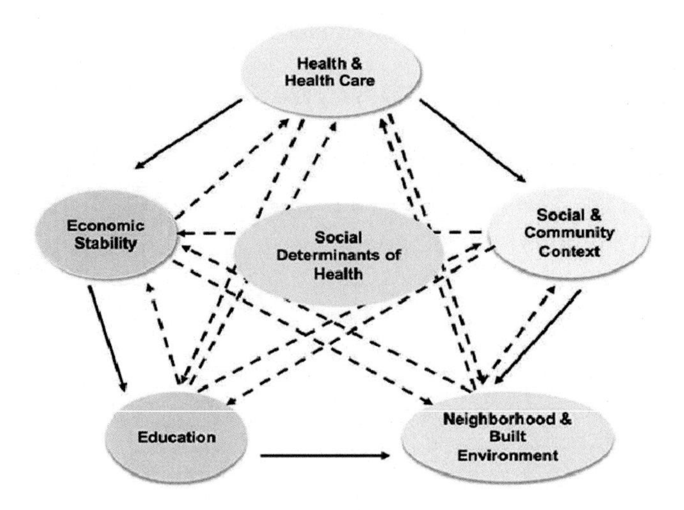

FIG. 6.10

How five domains of social determinants are overwhelmed with COVID-19 pandemic.

(ii) In COVID-19 contaminated cases, overall 20% of cases are severe or critical. The most victimized cases are of older age groups and in those with certain underlying conditions.

(iii) Besides, regular health-care system, social care systems, and measures taken to control transmission have had broad and deep socioeconomic consequences.

The overall clinical fatality in COVID-19 is currently over 3%, increasing with age and rising to approximately 15% or higher in patients over 80 years of age. CoVID-19 infected patients with previous health records such as cardiovascular, respiratory, and immune systems confront an increased risk of severe illness and death. Countries with early action and implementation of comprehensive public health measures such as rapid case identification, rapid testing and isolation of cases, comprehensive contact tracing, and quarantine of contacts have suppressed the spread of COVID-19 below the thresh at which health systems become unable to prevent excess mortality. Countries that have been a success in minimizing transmission and have under control the COVID-19 crisis could be able to minimize secondary mortality due to other causes through the continued safe delivery of essential health services.

In many countries, those who have been still suffering due to rapid transmission of COVID-19 are in the practice of keeping social distance to restrict the further spread of the pandemic disease. Commonly, physical distancing measures and restriction in movement, often referred to as "shut down" and "lock downs," can bring retardation in COVID-19 transmission. But, these measures are resulted in bringing social and economic life paralyzed. In addition, such measures disproportionately affect disadvantaged groups, including people in poverty, migrants, internally displaced people, and refugees, who most often live in overcrowded and under-resourced settings, and depend on daily labor for subsistence experienced mild disease, 40% will experience moderate disease including pneumonia, 15% of cases will experience severe disease, and 5% of cases will have a critical disease.

6.11.1 Rehabilitations during COVID-19

The past pandemic scenario of COVID-19 pandemic and present slow retardation process is resulted due to accurately diagnose and effectively isolate and intense care for all cases of COVID-19 including cases with mild or moderate disease. During the advanced stage of COVID-19, the primary attention was given to rapid identification, testing, and treatment of patients contaminated with COVID-19. In addition, temporary health-care shelters were developed to isolate the needy COVID-19 patients those who have poor housing facility. COVID-19 resulted in significant challenges for rehabilitation services around the world. Rehabilitation services are consistently among the health services most severely disrupted by the pandemic. At the same time, demand on rehabilitation of patients for temporary shelter has been increased in rate both for patients who are critically unwell with the disease and for those who continue to experience the long-term consequences of their illness. WHO, together with partners has issued guidance and resources for governments, health workers, and patients. This includes integrating rehabilitation into key clinical guidelines and developing standalone rehabilitation resources.

Patents, ventilated for a long time due to critical illness will usually require intense rehabilitation input as more than 60% are unable to walk. In case, a patient is experienced with a stroke or cardiac complications following COVID-19, he or she will require rehabilitation for an extended period, and in some cases, they will require lifelong support [51].

It is extremely necessary that patients are to be provided by health-care workforce [52], but during a pandemic when social distancing is required, health-care providers need to explore innovative approaches to rehabilitation to ensure that both COVID-19 and non-COVID-19 patients receive adequate care and support.

An inadequate rehabilitation service has significant impacts on patients, families, and health workers. Rehabilitation providers can adapt telehealth or virtual rehabilitation [53]. Telehealth refers to the delivery of health-care services via electronic communication (phone, internet, video calls) [54]. Telehealth can provide preventive; promotive and curative aspects of health and many different health-care professions are involved in its delivery [55].

6.12 World Organization for Disaster Risk Reduction

In The World Conference on Disaster Risk Reduction, held in Sendai, Japan, on March 14–18, 2015, the United Nations member's states adapted the international document "The Sendai Framework for Disaster Risk Reduction."

It is the outcome of stakeholder consultations initiated in March 2015, Supported by the United Nations Office for Disaster Risk Reduction at the request of the UN

General Assembly [56–58]. The Sendai Framework is the successor instrument to the Hyogo Framework for Action (HFA) 2005–2015, which had been the most encompassing international accord to date on disaster risk reduction. It is linked closely with the Sustainable Development Goals (SDGs) (UNGA 2015), Paris Climate agreement (UNFCCC 2015), and the outcomes of the World Humanitarian Summit [59] and Habitat III [60]. It is the successor agreement to the Hyogo Framework for Action (2005–2015), which had been the most encompassing international accord to date on disaster risk reduction.

The Sendai document framework resulted in the outcome of efforts made by the United Nations International Strategy for Disaster Reduction in which UN members of States, NGOs, and other stakeholders were also closely associated for an improved version of the existing Hyogo Framework, with a set of common standards, and a legally-based instrument for disaster risk reduction. It was also strongly proposed to tackle disaster reduction and climate change adaptation when setting sustainable development goals.

The overall aim has been to develop dynamic, local, preventive, and adaptive governance systems at the global, national, and local levels. In addition, the landmark UN agreements also aimed to agenda like universal health coverage, development, humanitarian action, disaster risk management, and climate change adaptation.

The Sendai Framework also plays a significant role in developing the traditional health profession to meet the urgency of disaster risk management across the global location in the grip of disaster. The basic logic of the Sendai Framework is to reduce and managing conditions of hazard, exposure, and vulnerability while building the capacity of communities and countries for prevention, preparedness, response, and recovery of the losses and impacts on health from disasters. This sort of attempt can be more effectively alleviated through a multisectoral approach rather than focusing exclusively on emergency response. Although there were many references to heath in the Hyogo Framework for Action 2005–2015 (UNDRR 2005), the adaptation of the Sendai Framework can show that the field of health and DRM has been substantially interwoven at the global multisectoral policy level [44, 61, 62].

Since the last couple of decades, the health sector has been performing dedicated service in developing evidence-based policy and practice, with globally recognized organizations dedicated to this endeavor. Followings are few such dedicated organizations work on resolving health risks problem resulted from the disaster.

6.12.1 Cochrane

Cochrane is one such leading example of such an organization. Cochrane (earlier known as the Cochrane Collaboration) is a British international Charitable organization that facilitates evidence-based choices about health interventions involving health professionals, patients, and policymakers [63, 64].

Cochrane is a nonprofit institution, set up in 1992 which works with over 130 contributors globally to produce high quality, accessible health information, and free from commercial interests. Cochrane claims that principles of evidence-based medicine are significant to health policy and practice, and can effectively work on disease prevention, treatment, and rehabilitation [65].

6.12.2 Evidence aid

Evidence Aid is another example of a leader in the field of evidence-based practice in the health sector. Evidence Aid works in collaboration with other organizations including Public Health England; Red Cross Flanders, International Rescue Committee; Centres for Disease Control; Centre for Evidence-Based Medicine; and the University of Oxford. The organization, which has charitable status in the United Kingdom, uses evidence from systematic reviews to provide up-to-date advice on interventions in the context of planning for or responding to disasters, humanitarian crises, and other major health-care emergencies [66].

6.12.3 Hyogo framework

The Hyogo Framework for Action (HFA) was the global blueprint for disaster risk reduction efforts between 2005 and 2015. Its goal was to substantially reduce disaster losses by 2015 in lives, and in the social, economic, and environmental assets of communities and countries. Hyogo Framework, mentions health as a sector and health-care facilities three times, but not as an explicit goal or outcome of disaster risk management (DRM).

6.12.4 The World Health Organization and partners

The World Health Organization and partnership with other voluntary organizations have a set of programs for managing and preventing health emergencies that predate the adaptation of the Sendai Framework.

The WHO Thematic Platform on Health emergency and Disaster Risk Management was launched by WHO and UNISDR in 2009 as an international, multiagency platform to advocate, share information, and catalyze action for Health EDRM. This resulted from 2008 to 2009 World Disaster Reduction Campaign on Hospital Safe from Disaster (UNGA 2008) and the 2009 Global Platform for Disaster Risk Reduction (UNGA 2009).

6.12.5 International Health Regulation (IHR) (2005)

International Health Regulations (IHRs) are a legally binding instrument of international law that was first adopted by the World Health Assembly in 1969 and last revised in 2005. It is mainly based on working out international collaboration "to prevent, protect against, control, and provide a public health response to the international spread of disease in way that are commensurate with and restricted to public health risks and that avoid unnecessary interference with international traffic and trade" [47, 67, 68].

The IHR is the only international legal treaty with the responsibility of empowering the World health Organization (WHO) to act as the main global surveillance system [48, 49].

In 2005, following the 2002–2004 SARS outbreak, several amendments were made in the IHRs originating from 1969. The 2005 IHR came into force in June 2007, with 196 binding countries that recognized that certain public health incidents, extending beyond disease, ought to be designated as a Public Health Emergency of International Concern (PHEIC), as they pose a significant global threat.

The credit for putting health risks and health resilience at the heart of global DRM goes to Sendai Framework which advocates for connecting health sectors for cooperating on emergency proactive and reactive measures globally, as well as highlighting the critical role of science and technology. The overall target of the Sendai Framework and IHR is to strengthen the resilience of the national health system through integrated disaster risk management into primary, secondary, and tertiary health care with strong emphasis at the community level; developing well-trained health workforce in understanding disaster risk and supporting and training community health groups in disaster risk reduction campaign.

The pre and post-disaster risk management planning should be made with the emphasis on life-threatening and chronic disease; to promote the resilience of new and existing critical infrastructure including water, transportation and telecommunications infrastructure, educational facilities, hospitals, and other health facilities, to ensure that they remain safe, effective, and operational during and after disasters.

Various organizations that have been supporting the DRM agenda are also involved in supporting Sendai Framework. Some of the organizations involved in the Sendai Framework, besides DRM activities are briefed below:

The World Association for Disaster and Emergency Medicine (WADEM)

WADEM is mainly associated with improving the delivery of prehospital and emergency care worldwide during disasters and other emergencies. It is one of the oldest international emergency and disaster medicine organization in the association of 55 countries and multiple disciplines, such as medicine, nursing, psychology, sociology, emergency management, and academia from both governmental and nongovernmental organization. In 2015 World Congress, the WADEM has adopted the Cape Town statement to extend support to the Sendai Framework and its implementation in the health emergency and disaster risk domain.

International Association of National Public Health Institutes (IANPHI)

The IANPHI was established in 2006 with the objectives to link, support, and strengthen the institute responsible for public health in countries worldwide, particularly in developing countries where poor facilities for health-care infrastructure is existing. The focus areas mainly include outbreak investigation and control, disease surveillance, health promotion, and emergency response.

The International Federation of Environmental Health (IFEH)

The IFEH is an umbrella organization whose members are the national environmental health organizations of countries, as well as universities and associated members. The IFEH is associated with 43 countries working in environmental health, mainly at local, regional, and state government levels.

The International Council of Nurses (ICN)

The International Council of Nurses (ICN) is a global federation of over 130 national nursing associations that endeavor to ensure quality nursing care universally, sound health policies, the development of nursing knowledge, competencies, and skills, and a respected nursing workforce. The main target of ICN is to work for the prevention of disaster risks management in association with volunteers working for disaster risk reduction. In 2019, the ICN updated their nurses and disaster risk reduction, response, and recovery policy statement in line with the Sendai Framework, highlighting that "nurse must be involved in the development and implementation of disaster risk reduction, response and recovery policies at the international level."

6.13 Conclusion

All countries, irrespective of their economic and social status, need perfect and well-planned strategies for the maintenance of sustainable and risk-free life even at the time of disaster situation either caused by natural or manmade. The governance of respective courtiers needs to have transparent policies, strategies, and related programs to mitigate health risks and their associated health consequences. The policies framed by the government should be multidisciplinary, intersectional, and compressive. In this connection, the role of the primary health-care center is supposed to be critical and most practicable to deal with Health EDRM recommended by international health risk organization and WHO.

References

[1] Plan of action for disaster risk reduction 2016–2021. Washington, DC: Pan American Health Organization, http://iris.paho.org/xmlui/handle/123456789/33772; 2016.

[2] Action plan to improve public health preparedness and response in the WHO European Region 2018–2023. Copenhagen: World Health Organization Regional Office for Europe, http://www.euro.who.int/_data/assets/pdf_file/0009/393705/Action-Plan_EN_WHO_web_2.pdf?ua=1; 2018.

[3] Resolution 70/1. Transforming our world: the 2030 agenda for sustainable development. Resolution adopted by the General Assembly on 25 September 2015. New York, NY: United Nations; 2015. A/RES/70/1 http://www.un.org/ga/search/view_doc.asp?symbol=A/RES/70/1&Lang=E.

[4] Sendai framework for disaster risk reduction 2015–2030. Geneva: United Nations Office for Disaster Risk Reduction, https://www.unisdr.org/files/43291_sendaiframeworkfordrren.pdf; 2015.

[5] International Health Regulations (2005). Geneva: World Health Organization. 3rd ed; 2016. https://apps.who.int/iris/bitstream/handle/10665/246107/9789241580496-eng.pdf.
[6] Paris Agreement. Bonn: United Nations Framework Convention on Climate Change; 2015. FCCC/CP/2015/10/Add.
[7] Aida J, Kawachi I, Subramanian SV, Kondo K. Disaster, social capital, and health. In: Kawachi I, Takao S, Subramanian SV, editors. Global perspectives on social capital and health. Berlin: Springer; 2013. p. 167–87.
[8] Aitsi-Selmi A, Murray V. Protecting the health and well-being of populations from disasters: health and health care in the Sendai Framework for disaster risk reduction 2015–2030. Prehosp Disaster Med 2016;31(1):74–8.
[9] Costello A, Abbas M, Allen A, Ball S, Bell S, Bellamy R, Kett M, et al. Managing the health effects of climate change: Lancet and University College London Institute for Global Health Commission. Lancet (London, England) 2009;373(9676):1693–733.
[10] Crawford L, Langston C, Bajracharya B. Participatory project management for improved disaster resilience. Int J Disaster Resil Built Environ 2013;4(3):317–33.
[11] Cutter SL, Ash KD. The geographies of community disaster resilience. Glob Environ Chang 2014;29:65–77.
[12] Cutter SL, Barnes L, Berry M, Burton C, Evans E, Tate E, Webb J. A place-based model for understanding community resilience to natural disasters. Glob Environ Chang 2008;18(4):598–606.
[13] Cutter SL, Burton CG, Emrich CT. Disaster resilience indicators for benchmarking baseline conditions. J Homel Secur Emerg Manag 2010;7(1).
[14] World disasters report 2016: Resilience: saving lives today, investing for tomorrow. Geneva: International Federation of Red Cross and Red Crescent Societies, https://www.ifrc.org/Global/Documents/Secretariat/201610/WDR%202016-FINAL_web.pdf; 2016.
[15] People affected by conflict-humanitarian needs in numbers. Brussels: Centre for Research on the Epidemiology of Disasters, https://reliefweb.int/report/world/people-affectedconflict-humanitarian-needs-numbers-2013; 2013.
[16] Disease outbreaks by year. Geneva: World Health Organization, http://www.who.int/csr/don/archive/year/en/.
[17] Fan VY, Jamison DT, Summers LH. Pandemic risk: how large are the expected losses? Bull World Health Organ 2018;96(2):129–34.
[18] World Bank, Institute for Health Metrics and Evaluation. The cost of air pollution: Strengthening the economic case for action. Washington, DC: The World Bank; 2016. https://openknowledge.worldbank.org/handle/10986/25013.
[19] Global assessment of national health sector emergency preparedness and response. Geneva: World Health Organization, http://www.who.int/hac/about/Global_survey_inside.pdf; 2008.
[20] Serrao-Neumann S, Crick F, Harman B, Schuch G, Choy DL. Maximising synergies between disaster risk reduction and climate change adaptation: potential enablers for improved planning outcomes. Environ Sci Policy 2015;50:46–61.
[21] Shaw R. Incorporating resilience of rural communities for proactive risk reduction in Shikoku, Japan. In: Kapucu N, Hawkins CV, Rivera FI, editors. Disaster resiliency: Interdisciplinary perspectives. New York: Routledge; 2013. p. 207–26.
[22] Sherrieb K, Norris FH, Galea S. Measuring capacities for community resilience. Soc Indic Res 2010;99(2):227–47.
[23] https://www.ifrc.org/Global/Documents/Secretariat/201610/WDR%202016-FINAL_web.pdf.

[24] World Disasters Report. Resilience: Saving lives today, investing for tomorrow, https://www.ifrc.org/Global/Documents/Secretariat/201610/WDR%202016-FINAL_web.pdf; 2016.
[25] https://www.who.int/csr/don/archive/year/2019/en/.
[26] https://www.worldbank.org/en/results/2017/12/01/climate-insurance.
[27] World Health Organization, http://www.who.int/healthsystems/about/en/ accessed June 2010.
[28] Dar O, Buckley EJ, Rokadiya S, Huda Q, Abrahams J. Integrating health into disaster risk reduction strategies: key considerations for success. Am J Public Health 2014;104(10):1811–6.
[29] Djalante R, Holley C, Thomalla F, Carnegie M. Pathways for adaptive and integrated disaster resilience. Nat Hazards 2013;69(3):2105–35.
[30] Drennan L, McGowan J, Tiernan A. Integrating recovery within a resilience framework: empirical insights and policy implications from regional Australia. Polit Gov 2016;4(4):74.
[31] Dufty N. Using social media to build community disaster resilience. Aust J Emergency Manag 2012;27(1):40.
[32] Field CB, Barros VR, Mach K, Mastrandrea M. Climate change 2014: Impacts, adaptation, and vulnerability. Cambridge and New York: Cambridge University Press; 2014.
[33] Kodner DL, Spreeuwenberg C. Integrated care: meaning, logic, applications, and implications—a discussion paper. Int J Integr Care 2002;2:14.
[34] Gröne O, Garcia-Barbero M. Trends in integrated care reflections on conceptual issues (EUR/02/5037864). Copenhagen, Denmark: World Health Organization; 2002.
[35] Leichsenring K. Developing integrated health and social care services for older persons in Europe. Int J Integr Care 2004;4. https://doi.org/10.5334/ijic.107, e10.
[36] International Health Regulations (2005) monitoring and evaluation framework. Geneva: World Health Organization, https://apps.who.int/iris/bitstream/handle/10665/276651/WHO-WHE-CPI-2018.51-eng.pdf?sequence=1; 2018.
[37] Bangkok Principles for the implementation of the health aspects of the Sendai Framework for Disaster Risk Reduction 2015–2030. Geneva: United Nations Office for Disaster Risk Reduction, http://www.who.int/hac/events/2016/Bangkok_Principles.pdf; 2016.
[38] Safe hospitals and health facilities. Geneva: World Health Organization, https://www.who.int/hac/techguidance/safehospitals/en/. [Accessed 31 March 2019].
[39] Smart hospitals. Washington, DC: Pan American Health Organization, https://www.paho.org/disasters/index.php?option=com_content&view=article&id=3660:hospitales-inteligentes&Itemid=911&lang=en.
[40] Busch N, Givens A. Achieving resilience in disaster management: the role of public-private partnerships. J Strateg Secur 2013;6(2):1–19. https://doi.org/10.5038/1944-0472.6.2.1.
[41] Cartalis C. Toward resilient cities—a review of definitions, challenges and prospects. Adv Build Energy Res 2014;8(2):259–66. https://doi.org/10.1080/17512549.2014.890533.
[42] Chamlee-Wright E, Storr VH. Expectations of Government's response to disaster. Public Choice 2010;144(1).
[43] Aitsi-Selmi A, Egawa S, Sasaki H, Wannous C, Murray V. The Sendai framework for disaster risk reduction: renewing the global commitment to People's resilience, health, and well-being. Int J Disaster Risk Sci 2015;6(2):164–76.
[44] Aitsi-Selmi A, Murray V. The Sendai framework: disaster risk reduction through a health Lens. Bull World Health Organ 2015;93(6):362.
[45] What are the International Health Regulations and Emergency Committees? www.who.int. World Health Organization. 19 December 2019.

[46] WHO EMRO. Background | About | International Health Regulations, www.emro.who.int; 2020. Retrieved 2 August.
[47] Heymann D. Part of Chapter 4. Public health, global governance, and the revised international health regulations. In: Infectious disease movement in a borderless world: Workshop summary. Washington, DC: National Academies Press; 2010. p. 180–95.
[48] Youde J. In: Youde J, editor. The International Health Regulations, Biopolitical Surveillance and Public Health in International Politics. New York: Palgrave Macmillan US; 2010. p. 147–75. https://doi.org/10.1057/9780230104785_7.
[49] Kohl KS, Arthur RR, O'Connor R, Fernandez J. Assessment of public health events through international health regulations, United States, 2007-2011. Emerg Infect Dis 2012;18(7):1047–53. https://doi.org/10.3201/eid1807.120231.
[50] WHO. COVID-19 strategy update - 14 April 2020, https://www.who.int; 2020.
[51] Faux SG, Eager K, Cameron ID, Poulos CJ. COVID-19: planning for the aftermath to manage the aftershocks. MJA; 2020.
[52] Bettger JP, Thoumi A, Marquevich V, De Groote W, Battistella LR, Imamura M, Ramos VD, Wang N, Dreinhoefer KE, Mangar A, Ghandi DB. COVID-19: maintaining essential rehabilitation services across the care continuum. BMJ Glob Health 2020;5(5), e002670.
[53] Dy Care. Rehabilitation and physiotherapy in times of pandemic. Available from, https://www.dycare.com/products/rehabilitation-and-physiotherapy-in-times-of-pandemic/. [Accessed 10 July 2020].
[54] Achenbach SJ. Telemedicine: benefits, challenges, and its great potential. Heal Law Pol Brief 2020;14(1). Available at: https://digitalcommons.wcl.american.edu/hlp/vol14/iss1/2.
[55] Cottrell M, Russel T. Introduction to telehealth course. Phys Ther 2020.
[56] Rowling M. New global disaster plan sets targets to curb risk, losses. Reuters; 2015.
[57] Sendai. A new global agreement on disaster risk reduction. Overseas Development Institute; 2015.
[58] Many Disaster-related Meetings, Exhibitions to be Held. The Japan Times, https://www.japantimes.co.jp/news/2015/03/14/national/many-disaster-related-meetings-exhibitions-held/#.VdH7krfhlKj.
[59] UNGA (United Nations General Assembly). Outcome of the World Humanitarian Summit: Report of the Secretary-General (A/71/353 2), https://www.agendaforhumanity.org/sites/default/files/A-71-353%20-%20SG%20Report%20on%20the%20Outcome%20of%20the%20WHS.pdf; 2016.
[60] UNGA (United Nations General Assembly). New Urban Agenda, http://habitat3.org/the-new-urban-agenda; 2016.
[61] Aitsi-Selmi A, Egawa S, Sasaki H, Wannous C, Murray V. The Sendai framework for disaster risk reduction: renewing the global commitment to people's resilience, health, and wellbeing. Int J Disaster Risk Sci 2015;6(2):164–76. https://doi.org/10.1007/s13753-015-0050-9.
[62] Maini R, Clarke L, Blanchard K, Murray V. The Sendai framework for disaster risk reduction. Science 2017;8(2):150–5. https://doi.org/10.1007/s13753-017-0120-2.
[63] Public Health Guidelines, https://www.nihlibrary.nih.gov/resources/subject-guides/evidence-based-public-health/public-health-guidelines.
[64] Hill GB. Archie Cochrane and his legacy. An internal challenge to physicians' autonomy? J Clin Epidemiol 2000;53(12):1189–92.
[65] Turner T, Green S, Tovey D, McDonald S, Soares-Weiser K, Pestridge C, Elliott J, on behalf of the Project Transform Team, IKMD developers. Producing Cochrane systematic

reviews—a qualitative study of current approaches and opportunities for innovation and improvement. Syst Rev 2017;6(1), 147.

[66] Khalid AF, Lavis JN, El-Jardali F, et al. Stakeholders' experiences with the evidence aid website to support 'real-time' use of research evidence to inform decision-making in crisis zones: a user testing study. Health Res Policy Sys 2019;17:106. https://doi.org/10.1186/s12961-019-0498-y.

[67] WHO EMRO. Background | About | International Health Regulations. Retrieved 2 July http://www.emro.who.int/international-health-regulations/about/background.html; 2021.

[68] What are the International Health Regulations and Emergency Committees? World Health Organization. 2 July www.who.int; 2021.

CHAPTER

Health-care marketing and supply chain management 7

7.1 Introduction

Timely supply of medicines and health-care aids from manufacturers to the consumer through a series of systematic logistic systems with sustainable, safe, and secure management is a challenging issue. So health-care supply chain management (SCM) is a complex process involved in the collection of health-care aids and medicines, monitoring of the supply process, and delivering goods and services to providers and patients (Fig. 7.1). The overall process of supply chain of health-care products passes through a number of independent stakeholders, including manufacturers, insurance companies, hospitals, providers, group purchasing organizations, and several regulatory agencies. By developing innovative technology, health-care organizations can overcome major challenges to further reduce spending. Supply chain management starts with the procurement and distribution of products and services as they move from the receiving dock to the patient. But during such a transition, SCM confronts major issues including the hoarding of supplies, demand for specific types of products in inventory, product expirations, out of stock issues that may lead to expensive delivery charges, pilferage, and unwanted increase in inventory dollar-based demand. These and other issues contribute to out-of-budget supply cost.

7.2 Supply chain management history

In the early 1950s, the manufacturers and promoters could realize the benefit of quality control in business for durable and work well commodities in the common market. Mainly, this practice was used to manage through one promoter to minimize the production expenses. The promoters relied on partnership business. The overall management of business starting from production to the wholesale or retail selling process used to be under control through self-managed logistics system. The practice of keeping large inventories in store was supposed to be helpful in the continuous manufacturing process. Subsequently, due to tough competition, in the beginning of the 1970s, the promoters started realizing that minimizing the quantity of inventories would be helpful in reducing the overall expenses. Gradually, the business community started realizing the importance of integration of functional aspects of stockholders for the future prospects of sustainable supply chain management in the global market.

Healthcare Strategies and Planning for Social Inclusion and Development. https://doi.org/10.1016/B978-0-323-90446-9.00007-1

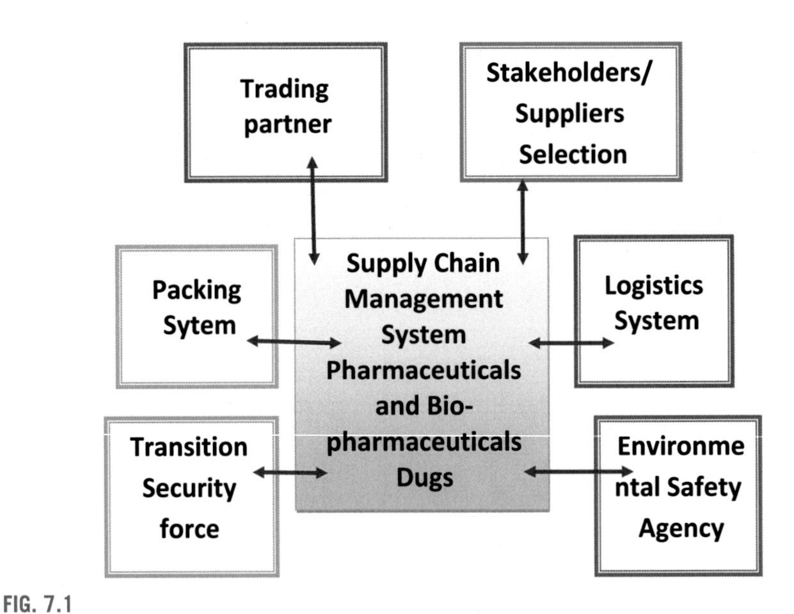

FIG. 7.1

Supply chain management system integrated with different stakeholders.

So the promoters of companies started to follow material requirement planning (MRP) and manufacturing resources planning (MRP II) systems to minimize inventory holding. In due time, the promoters started realizing the benefits of MRP and MRP II. This could only be possible by the application of highly sophisticated software packages (LAN and WAN) based on information technology and data analysis. This process started giving updated information not only on tracking the goods in the logistics process to know about the inventories condition, but also on goods transition. This could have been possible by the use of just-in-time (JIT) and total-quality-management (TQM) software packages.

During the 1980s, the business front was well equipped with the management technology on systematic process feedstock procuring methods, integrating the approach in the manufacturing management process and well-improved logistic system for timely delivery of goods with safety and maximum security. In addition, the manufacturers acquired control over quality product development and quality assurance in order to develop confidence among the end users. In 1919, the basic concept on temperature-controlled logistics systems was used by Parke, Davies & Co (later branded as Pfizer) for transport of biopharmaceuticals through an effective and highly efficient supply chain management process. Subsequently, biopharmaceutical manufacturers started using a thermoregulated logistics system for the distribution of biologic drugs such as therapeutic proteins, blood factors, thrombolytic agents (tissue plasminogen activator), hormones (insulin, glucagon, growth hormones, gonadotropins), hematopoietic growth factors (erythropoietin, colony stimulating factors), interferon (interferon-α, $-\beta$, $-\gamma$), etc., as these medicines are gaining demand for the treatment of life-threatening diseases like cancer, autoimmune disease, anemia, etc. Since then, supply chain management has become an integral part in bringing sustainability in the business operation process.

Since then, the consumer's attention has increasingly diverted toward the introduction of marketing oriented branded products. During the Industrial Age of the 18th and 19th centuries, quality product oriented industries/companies could thrive due to the scarcity and high demand for mass-production, high-quality commodities and services. Industrialists/business people have started exploiting the possibilities of economies of scale to reach maximum efficiency at the lowest cost. With the emergence of sophisticated information technology and availability of a wide range of data banks, supply chain management has undergone a sea change in the 21st century. Although the pharmaceutical supply chain is a small fragment of the other goods supply chain, it plays a significant role in the health-care management system, especially during the emergency period caused due to natural disasters.

The emerging technology on GS1 standard and unique device identification (UDI) for medical devices is expected to boost health-care marketing in the coming years. Global Trade Item Number (GTIN) can be used by a company to uniquely identify all of its trade items. GS1 defines trade items as products or services that are priced, ordered, or invoiced at any point in the supply chain. So global standards such as bar codes have significantly improved patient safety and supply chain efficiency and effectiveness in health-care marketing management. In addition, GS1 can be helpful in serialization, authentication, and traceability to catch duplication and unauthorized serial numbers. According to the WHO, about 15% of the world's medicines are counterfeit. So the stakeholders can verify the supply chain history for each product.

World regulatory bodies such as the US Food and Drug Administration (FDA), the European Commission have made it mandatory for marketing and supply chain for quality control and services of medical devices, by the implementation of IDI.

Presently, cloud-based supply chain management software and its applications are transforming the health-care industry. Cloud computing allows marketers to store important data in a global network that can be accessed from anywhere in the world and from any device. This in turn allows marketers in the pharmaceutical industry to function more efficiently where they are, instead of having to be in the office to access specific data.

For instance, Jump Technologies has developed a cloud-based software solution called Jump Stock that integrates with EMR, ERP, or scheduling systems, thereby allowing hospitals to reduce costs associated with supply hoarding, physician preference variances, and stock-outs. Cloud-based track and traceability solutions protect manufacturers from variable expenses of a product recall by early identification of product quality problems. These solutions also improve order management and production planning across all warehouse locations. Cloud-based Jump Tech provides the best solutions for inventory management for hospitals, which are easy to implement, adopt, and manage using mobile devices such as tablets and smart phones.

7.3 Health-care marketing

Health-care marketing is related to giving a boost to pharmaceuticals and biologic drugs, marketing the services of physicians, and educating the public about the role of the health professionals in their communities, which are some of the important aspects of health care for all. The continuous climatic changes, natural disasters, and modern fast-food habit have been increasing the demand for safe and secure health-care aids and pharmaceuticals on time. So it is necessary to develop and manage a sustainable supply chain management system to meet the global pharmaceutical market. The American Hospital Association held its first marketing convention in 1977. It was related to highlighting the collaborative effort focused on promoting various aspects of the health industry. Marketing in health care has since become big business.

The global market of health-care products is in an emerging state with the opportunity of jobs both in developed and developing countries. It has been predicted that the health-care market will increase threefold to US$ 133.44 billion by the end of 2022. Health-care marketing includes health care which comprises hospitals, medical devices, clinical trials, outsourcing, telemedicine, medical tourism, health insurance, and medical equipment. In 2020, global advertising for the health-care industry in the United States was about 5%, turning it into a $36 billion industry. The Chinese health-care industry is currently ranked the second largest in the world behind the United States. In 2019, the Chinese market reached RMB 7.82 trillion (US$ 1.1 trillion). The world's second biggest populated country, India's health-care sector is growing at a brisk pace due to its strengthening coverage, services, and increasing expenditure by public and private players.

The overall response of health-care marketing in the global market depends on the growing and aging population, rising prevalence of chronic diseases, infrastructure investments, technological advancements, evolving care models, higher labor costs amid the workforce shortages, and the expansion of the health-care system in the developing market.

Advanced IT technology has led the developed countries in health-care marketing enormously (Fig. 7.2).

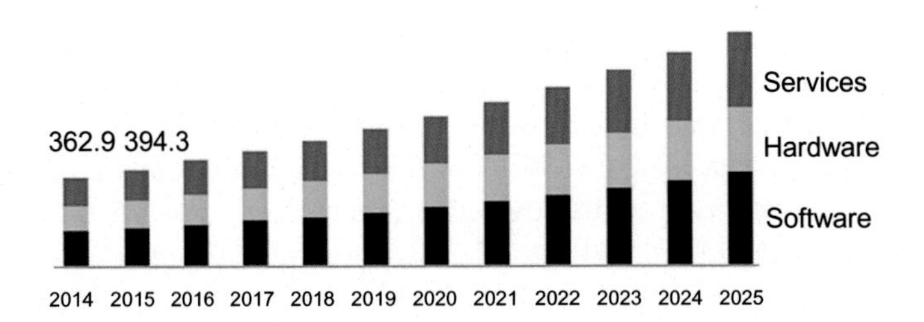

FIG. 7.2

US health-care supply chain management market size, by product, 2014–2025.

Source: www.grandviewresearch.com.

The main problem with biopharmaceutical companies is how to reduce the cost of the supply chain of biologic drugs that are perishable at a higher temperature. So it is necessary for pharmaceutical and biopharmaceutical companies to have an option for cryologistic carriers in the supply chain for biologic drugs. Any restriction in the implementation of the supply chain may adversely create problems in bringing sustainability in health-care marketing. In addition, the health-care supply chain is mainly based on two operational systems, software and hardware. The software system is highly acceptable compared to the hardware system due to the increasing number of online purchases, improving business intelligence, and growing preference for green logistics.

On the basis of end users, the overall supply chain management system is divided into: receiving quality material for production, the manufacturing system, product packing and storage, distribution through a suitable logistic system (Fig. 7.1).

The highly complex molecules derived from biological systems are recognized and appreciated due to their body friendly nature. Due to their high-temperature sensitivity, biologic drugs require intensive care in handling and transporting to the respective end users. So the logistics system for biologic drugs needs an integrated supply chain management process that is controlled by the latest IT aids. Still, the marketing of biopharmaceuticals faces big challenges such as high cost, some side effects like breathing problems, sore throat, allergic reactions, itching, and rashes while consuming biologically derived therapeutic drugs. The design and discovery of low-cost biosimilars has been challenging the sales of biologic drugs due to the lesser cost of biosimilars. In the true sense, biosimilars are the generic form of biologic drugs (after their expiry of patent time) having an equivalent potential in fictional as noticed in biologics. Biosimilars are also approved by the FDA for the equivalent properties possessed by biologic drugs. In spite of the many technological challenges and disadvantages, biopharmaceutical companies have been in a rising stage.

The health-care industry is a highly integrated system to facilitate goods and services to patients, even at the time of crises like natural disasters or any other activities causing an alarming condition at the community level, both for developing and underdeveloped countries to reestablish a sustainable pattern of health.

Although health-care marketing is a small fragment of the total hardware and other commodities market, it is the world's largest growing industry covering about 10% of the gross domestic product (GDP) of most developed nations. Health-care marketing can be broadly divided into two categories: marketing and supply of health-care aids, and supply chain management of pharmaceutical and biopharmaceutical drugs.

Marketing and supply of health-care aids

Health-care aids consist of equipment and services groups of companies and entities that supply medical equipment and health-care services which include:

- Labor time of various trained professionals, specialists, nurses, medical technicians, pharmacists, and many others
- Procedures and testing, such as magnetic resonance imaging (MRI) scans and laboratory analyses of blood samples

- Hospital and nursing care services
- Emergency services such as ambulances
- Pharmaceutical products (bandages to chemotherapy drugs)

Supply chain management of pharmaceutical and biopharmaceutical drugs

In this category, the listed industry group includes companies that produce biotechnology, pharmaceuticals, and miscellaneous scientific services. Due to the increase in complexity of manufacturing, pharmaceuticals and medical devices companies are keen to keep updated information on the global marketing status and half lifecycle of products. Currently, it has been more challenging for the pharmaceuticals and health-care companies to keep sustainability in quality and compliance issues. Still, health-care marketing and supply chain management are a small fragment and incomplete, which may bring risk to patients and hospital management authorities.

7.4 Role of cold chain logistics in pharmaceuticals

Health-care marketing mainly consists of two types of goods: (i) health-care aids such as equipment, instruments, and other nonprescribed items, (ii) prescribed medicines such as pharmaceuticals (generic in nature), biopharmaceuticals such as therapeutic proteins, and other biologic (biosimilar, biobetter) drugs.

The former category of health-care items can be supplied through ordinary logistics without any provision of a temperature regulation system. The later types of items consisting of different types of biopharmaceuticals and other biologics need cold logistics for delivering to end users on time and safely without any harm to the quality of products. So it is necessary to ensure product integrity by quality control and quality assurance before handing over pharmaceutical and biopharmaceutical therapeutic products to the end users. The cold chain logistics system is only an option for such a transition by which a biological origin product can be safely transferred from the manufacturing site to the end users. The transfer of materials in the cold chain system involves a fully temperature controlling equipment and instruments which can provide a guarantee of maintenance of the desired temperature as instructed by the manufacturer. The cold chain system is a highly integrated in which information technology plays a significant role in keeping contact with the movement of product and the manufacturer in order to attain sustainability in the manufacturing process.

Global regulations on the cold chain management system and operational control are highly strict. Both the manufacturer and stockholders cum retail dealers want to receive proof of data on instant temperature as per the necessity of the product under transfer process. Generally, a single consignment involves high value. So it is necessary to have a proper insurance policy to meet the damage of product due to unwanted incidents during the logistics operation process.

The overall process of the supply chain management system is involved with science, technology, and process management. It is a science because it needs an understanding of the biological characteristics of therapeutic molecules and their

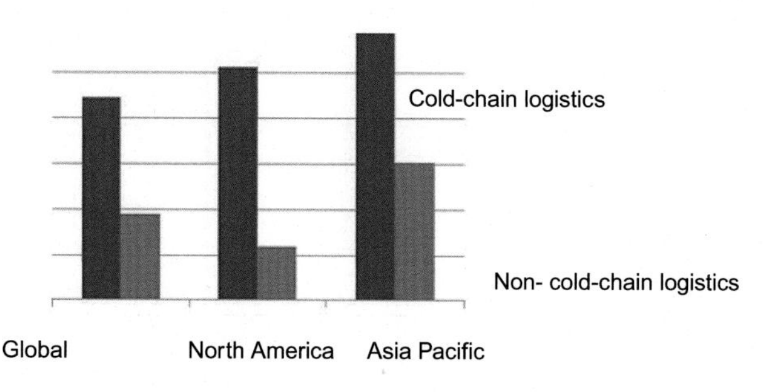

FIG. 7.3

Biopharmaceutical market logistics growth rate 2017–22%.

Source: The Business Research Company; Market Research.com.

interaction with the surrounding microenvironment. It is known as a technology because it needs a dedicated temperature controlling system based on a microchip operation system. It is a process due to the involvement of a sequential transfer process starting from the site of production to reach retail dealers for further process in transferring the product to the end users in a safe and secure manner. The cold chain logistics market, which moves biologics, is growing faster globally than the noncold chain logistics market which transports small molecule drugs. Between 2012 and 2018, the cold chain logistics market more than doubled in size, while the noncold chain market grew by only 36% in comparison (Fig. 7.3).

7.4.1 Facility growth in health-care marketing

Generally, the third-party logistic service for biopharmaceutical products is mainly based on microchip linked integrated operational management, provision of temperature controlled warehouse facilities, and global linked network facilities for timely distribution among retail dealers. In addition, third-party logistics gives a guarantee for delivering goods on demand basis. In some cases, as a special service, the third party extends the service on the customer's requirement based on market conditions such as demand and delivery service requirements on the basis of product quality and quantity. A third party having these facilities is known as a third-party supply chain management provider or as a supply chain management service provider (SCMSP).

With advanced information technology, the emergence of third-party logistics (3PL or TPL) has stated playing a significant role in catalyzing the service on supply chain based on quality. Mostly, third-party logistics (3PL) gives special services on warehouse transportation and outsourcing of raw material provision [1].

7.4.2 Clinical logistics for biopharmaceuticals

Due to the highly competitive market of biopharmaceutical products, it has been a challenging operation for clinical logistics management. Due to the implementation of a strict regulatory act, clinical logistics has become a complex and committed

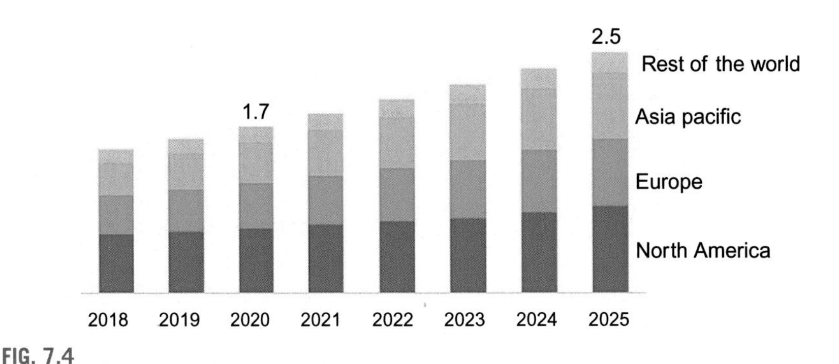

FIG. 7.4

Clinical trial supplies market, by region (USD Billion).

Source: marketsandmarkets.com.

task while delivering trial-related supplies to multiple locations around the world. The global clinical trial supply and logistics market is expected to grow from an estimated $14.84 billion in 2016 to $28.75 billion in 2027 [2]. The global clinical supplies have been gaining continuously increasing orders and may reach about 2.5 billion USD at the end of 2025 (Fig. 7.4).

Sample analysis and in vitro studies related to the clinical impact of biologic drugs on human trials for various localities need intense care and timely delivery of materials on the spot. So it has become a part of the duty of clinical logistics to extend services with all the necessary laboratory services and ancillary supplies for clients. The complete package for ancillary supplies consists of testing kits related to clinical trials on the spot, small and easily portable diagnostic equipment, relevant protocol for immediate reference and follow-up. Nowadays, it has become mandatory for a clinical trial agency to provide the necessary drug requirements for the patients under trial by coordinating the supply chain process management between the manufacturer, distributor, and end user. In addition, it is the duty and moral obligation of the clinical trial agency to dispose of leftover waste materials safely and securely. Onsite laboratory management includes distributing the data collection sheet among the persons under clinical trial and submitting to the doctor after preliminary data compliance, ensuring coordination between the centralized laboratory and temporary laboratory for field studies.

Following are some of the significant advantages of clinical logistics:

- It brings coordination between global leading agencies dealing with logistic clinical jobs.
- It generates integrated data analysis based on global regulatory expertise.
- It finds technology solutions on the database results for its future use.
- It develops integrated planning for the clinical trial process for effective and time-bound results.
- It reduces overall expenses on clinical trials, this is taken at different locations around the world.

7.4.3 Global status and trends of biologics

The global biopharmaceutical market has been classified on the basis of types of biologics and applications: monoclonal antibody, interferon, insulin, growth, erythropoietin, growth coagulation factor, vaccine, hormone, and others (Fig. 7.5). By application, it is categorized into oncology, blood disorder, metabolic disease, infectious disease, cardiovascular disease, neurological disease, immunology, and others.

Biologically derived materials used in clinical trials are temperature sensitive and need intensive care in handling during logistic services. Generally, biologic drugs are produced and packed under regulated temperature conditions. After packing, the packed product is processed for shipping under temperature-controlled logistic systems with the provision of digital tracking devices. Some pharmaceutical products such as blood samples, tissue-derived biomaterials, and clinical trial materials need special supervision. The biologically derived drugs are highly cost oriented. However, due to minimum side effect, people prefer to use biologically derived medicines. But the body friendly nature of biologics has a high rate of demand.

Application of microchip technology in supply chain management has brought a tremendous revolution in timely distribution of biopharmaceutical products to the different locations of the world. The introduction of a digital IT-based operation system and tracking of products during logistics operation conditions have not only developed the practice of timely delivery but also taken care of product quality management till it reaches the end users. The major top biopharmaceutical companies have an in-house SCM team to take care of product packing to manage efficient green logistics systems. All the products in transition of the supply chain are well packed. The management of quality of the packed material mainly depends on the efficiency of the packing technology. So it is highly necessary to keep in touch with physical logistics flows in order to ensure safety delivery process with the guarantee

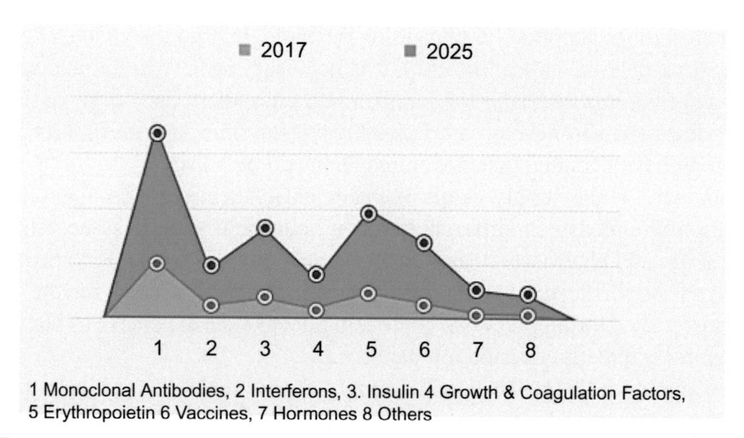

FIG. 7.5

Global biopharmaceutical market.

Source: www.alliedmarketresearch.com.

of product quality. Nowadays, companies are particular about packaging characteristics (e.g., physical status of packing size, shape, and material used for this purpose) in order to improve the image of the companies and minimize their cost. Recently, the emergence of electronic commerce or e-commerce has proven to be immensely beneficial in the supply chain management process and timely delivery of goods to the end user. So the manufacturers are having well-organized groups for packing, freight forwarding and transportation, and electronic monitoring of both a packing's internal conditions and process followed by carriers, to meet the requirement of industries for sustainable growth and development.

The frozen and cryogenic biologics are transferred in small quantities, but need a cold chain logistic system for transfer from the destination of manufacture to wholesale and retail dealers. The cryogenic transport systems are highly equipped with temperature control equipment and instruments. Often, the liquid nitrogen circulation system (below − 150°C) is provided inside the carrier vehicles to carry cord blood stem cell products safely. So the cost of cryogenic logistics is highly expensive, as compared to ordinary transport systems.

7.4.4 Expansion of health-care marketing

Marketing and its expansion play an important role in helping health-care professionals to create, communicate, and provide value to their target market. Modern marketers start from customers rather than from products and services. They are more inclined toward developing a sustainable relationship, rather than in ensuring a single transaction.

In addition, continuous increase in patent expiration, and increase demand on biologic drugs, and realization on the body friendly nature of biologics, the pharmaceutical industries are in extreme dilemma whether to continue further with chemically derived generic medicines or to diversify into biopharmaceuticals. In order to go in for an additional option for expanding the biologic drug manufacturing process in the existing pharmaceutical industry, it is necessary to have additional space and develop a special infrastructure for sophisticated equipment operation. Besides this, the companies have to develop a special logistics system and third-party logistics operation with well-trained manpower and technical persons.

As compare to the supply chain management of heavy goods like automotive parts, agriculture feedstock, different types of household materials, the value of the biologic drugs and pharmaceutical products supply chain is supposed to be insignificant. But the sophisticated temperature controlled logistic system for the biologic drugs transfer and storing process is more complicated and expensive as compared to other inanimate material distribution patterns.

For example, in 2015, the FeedEx supply chain added 1.1 million square-foot multitenant distribution warehouses on East Holmes Road. Memphis-based FedEx Corp. acquired Genco, a third-party logistics provider, for $ 1.4 billion.

In this connection, another good example is the expansion pattern of UPS Health-care Logistics (which operates more than 60 facilities for health-care

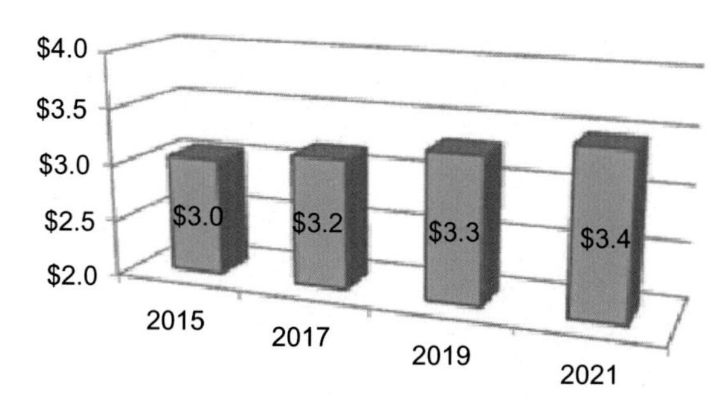

FIG. 7.6

Clinical trial logistics continues a gradual upward climb.

products worldwide), which developed a second facility near Bogota, Colombia with huge space. Presently, UPS Health-care Logistic has approximately 8 million square feet of cGMP or cGDP compliant health-care distribution space. UPS also possesses the world's largest network of field stocking locations with about 900 sites. It has been observed that with the increase in clinical trials of biopharmaceuticals, the logistics expenses are in increasing order at the global level (Fig. 7.6).

7.4.5 Cold chain regulation

The cold chain is a logistics management process for products that require the refrigerated temperature that customers demand. The supply of pharmaceuticals and biologics heavily rely on the cold chain in order to the quality and half lifecycle of the goods.

Biologic drugs are extremely prone to elevated ambient temperature. So the supply chain management for biological products is well equipped with a temperature controlling system with instant tracking provision.

Thus, a biopharmaceutical supply chain management system gives assurance for discharge of goods under controlled temperature conditions as desired by the manufacturer. Some vaccines for COVID-19 need extremely low temperature (−70°) for storage soon after manufacture. For example, in 2020, Pfizer developed the provision for storing finish COVID-19 vaccines in an ultracold storage system, at −70°C, until the Food and Drug Administration approves use and the vaccines can be distributed (Fig. 7.7).

The federal regulations related to the cold chain management process and application explain about the food quality and safety but nothing on biopharmaceutical products management. The FDA rules explain and ensure food safety from farm to consumer. The Hazard Analysis and Critical Control Points (HACCP) addresses food safety by controlling hazards from production to consumption, but nothing is mentioned about biopharmaceutical products safety and security during distribution to

FIG. 7.7

Photograph of ultrastorage, at −70°, developed by Pfizer.

retail stockists and end users. Only focus is given to dairy, juice, and sea foods from production to consumption.

Presently, in order to meet the demand for biopharmaceuticals in the world market, the manufacturers prefer to avail a temperature controlled logistic system having well-equipped instruments to provide information on the condition of shipping materials (blood sample, tissue, blood serum, stem cells) instantaneously (Fig. 7.8).

In addition, biopharmaceutical manufacturers want an assurance on the protection of biological origin materials through tracking microchips attached to the containers. Additional care is taken by the cold supplier for the phase III trial of biological materials by recording the variable temperature of the inside container, which may be helpful while assuring the quality sample for safe investigation and analysis.

7.4.6 Cold chain logistics providers

The cold chain logistics system for biopharmaceuticals is much smaller as compared to global logistics. However, the development of new cooling technology, implementation of strict Regulatory Acts, third-party logistics are the major factors that have brought in complexity and cost intensive in cold chain logistics for biologics. Several of the major global air cargoes are having a well-managed third party with logistics temperature controlled inventories storing system and tracking facilities. In addition, logistics providers and freight forwarders also have good network base control towers having well-trained staff to monitor shipments by assuring safe delivery.

7.4.7 Engineering of packaging

Pharmaceutical companies have well-trained packing engineers to design active and passive containers for the safety and security of biological origin materials in the transition stage. The active containers are well designed for maintenance of

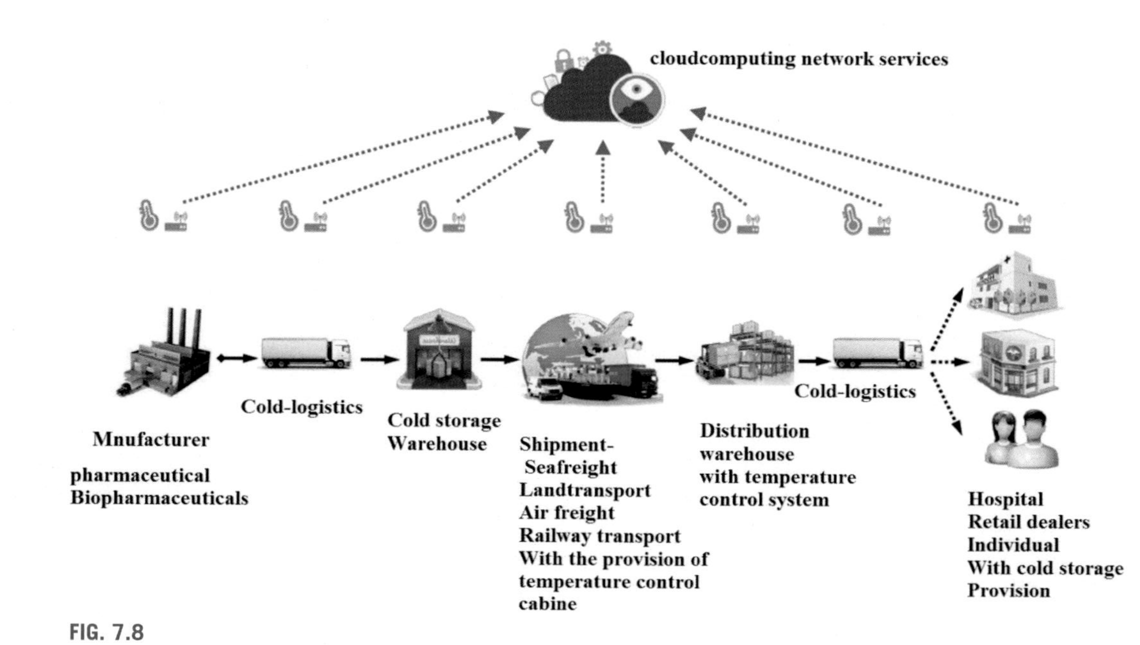

FIG. 7.8

Cold chain supply system for transport and distribution from manufacturer to end users.

cryogenic temperature, and the passive containers generally contain a set of chilling materials to provide cooling. The active containers are designed to maintain cryogenic temperature, even when the flight or trucking schedules are upset. The active containers are generally reused after proper cleaning. But the passive containers are rejected after use. The overall cost of a shipment is the balance between the cost of the containers, the cost to ship the container, and return-logistics or disposal costs.

The thermal blanketing packing system is followed by the entire European community as a good distribution practice (GDP). Thermal blanketing is not restricted to biopharmaceutical products but is widely applicable to controlled room temperature (CRT) shipments under 15°C–25°C. In this connection, shippers follow strict GDP rules in order to keep CRT shipments within the temperature range claimed on the drug label, or provide good reasons related to the stability of the product to be shipped.

In the initial stages, DuPont used to supply nonwoven plastic sheeting as thermal blanketing packing material. But at a later stage, DuPont started manufacturing a third-generation product, branded as Xtreme WD-50, to avoid accumulation of condensed humidity inside the packing container. The first-generation thermal sheet is simply a Tyvek cover to protect against solar heating and, to some extent, controlling airflow. The second-generation sheet is a multilayer product that includes an insulation layer to enhance temperature control. Now the third-generation product provides for reflectivity, insulation, and vapor control.

7.4.8 Managing supply chains from the cloud

Modern information technology (IT) has catalyzed the efficacy in operation of the supply chain management and logistics system. Introduction of advanced IT packages has helped the manufacturers in doing collaborative network communication all over the world for the customized-based supply chain operation and timely availability of product to the end users [3]. A variety of IT-based tools, i.e., shipping status tools, order processing tools, lean inventory tools, warehouse management tools, supplier management tools; demand forecasting tools, analytics and reports tools, security feature tools, compliance tools, etc. are presently being used for the high-level collaboration of business assignments across the world. By the implementation of IT tools and techniques, Toyota was able to increase the overall output to 140-fold, cause a 25% reduction in inventory and other benefits [4]. So it is high time the top manufacturers integrated their resources with IT tools which may be helpful in tightening the business collaboration among key stakeholders for the further development of the supply chain management process. The last decade has witnessed the benefit of cloud computing tools and techniques in developing world network facilities for the further prosperity of business.

This tool can be used for the supply chain management of a variety of materials including biopharmaceuticals, multiple business processes, and third-party logistics operation. Implementation of cloud computing tools and techniques is low cost and easy to handle with basic primary knowledge on the computer operation system.

Business partnership has the advantages of using IT tools as a common package as partners can easily implement cloud-based applications instead of purchasing and installing expensive software, allowing organizations to work together faster.

7.5 Risk management of the supply chain

Due to the temperature sensitivity of biopharmaceuticals, the supply chain management process is not only challengeable but also risky. So risk management is an integral part of the supply chain operation system. In order to have a risk free successful business, a company should be free of a litigated management system with the implementation of regulatory guidelines. In addition, well-managed QC and QA are integrated parts to bring acceptability of biological drugs in the international market without any business risk. The practice of risk management increases the level of confidence in members of a company but ensures that the end user gets the products on time. So it is necessary to implement risk management practice in the manufacturing process and transfer of product to a third-party logistics agency.

Product safety, security, consistency, and quality supply are the major challenges to SCM and logistics operations. So risk management plays a critical role while transferring the products from the site of manufacture to retail dealers. Both the United States and Europe follow ICH Q9 guidelines for SCM and logistics processes. Microbial contamination is a major risk factor in the biopharmaceutical production industry. Biopharmaceuticals get contaminated mostly during hold times, either at the processing stage or as inventory in a storing place. The primary factors that affect the risk of microbial contamination during hold times are due to the growth promoting properties of the in-process materials, initial bioburden level, and storage conditions. It has been noticed that microbial contamination is the most prevailing factor for the failure of risk management. In order to ensure risk free manufacturing, FMEA (failure mode and effects analysis) should be sincerely implemented, even with suitable modification, if necessary. The continuous supply of product components on quality base for a manufacturing process is a key factor to avoid risk in the manufacturing process. So raw material control strategies play a crucial role. Selection of honest persons, solvency, duration of business, and vendors' in-house quality system are important issues and should be taken care of while negotiating business strategies with vendors. The vendors should have their business continuity plans (BCPs) as per international standards.

7.6 Significance of logistics operations

In real practice, the overall growth and development of business is assessed on the basis of wide variable activities. As an example, supply chain management is a vital part of business management to maintain sustainability in the manufacturing process and to compete with other equivalent business in a successful manner. In practice, the logistics and supply chain management are complementary to each other and play a significant role in overall business management. So logistics management has highly significance value to have a successful business in sustainable pattern. At present, supply chain management is being strictly regulated by internationally regulatory bodies like the FDA in order to bring security and safety in delivering biologic drugs and other related materials to retail dealers located at different geographical areas on time. So, it is advisable for the promoters of biopharmaceutical companies to integrate modern green supply chain and logistics management to compete with other related companies like channel management, value chain management, and value stream management [5–7]. Generally, in the supply chain management system, being competitive with a counterpartner is more important than getting more profit. It is mainly due to the involvement of multiple organizations instead of a single company. Thus, a supply chain represents a cluster of organizations instead of a single unit [8, 9]. Generally, for biopharmaceutical companies, time is a major concerning factor due to loss of patent life, competition with generic drugs, and mounting economic pressure. The logistics of biopharmaceuticals usually involves the integration of information flow, material handling, production, packaging, inventory, transportation, warehousing, and often security. The complexity of logistics can be modeled, analyzed, visualized, and optimized by dedicated simulation software. Minimization of the use of resources is a common motivation in logistics for import and export.

7.7 Supply chain operation

Order management with multiple parties can be well managed by means of a Real-Time Value Network (RTVN) (Fig. 7.9).

Through this process a buyer company purchase order and the selling company can sale order in a single business transition. It is an integrated process with the involvement of buyers, sellers, movers, shippers, and service providers for the efficient transition of goods from the manufacturer to wholesale dealers. This type of business transition is helpful in increasing the efficacy of a specific business, in a cost-effective manner and in the shortest possible way.

In the traditional process, goods transition takes an uncertain time to fulfill the sale of a targeted quantity of product in a scheduled time. RTVN is an innovative method for business planning to fulfill the demand and supply of multiple transition of different locality with a single attempt. The RTVN business protocol is a fair way of conducting a business practice without any litigation. In addition, the RTVN methodology provides a complete package of business transition, starting from sales orders and purchases orders, to claims and returns all in a single system accessible anywhere. RTVN can also be helpful in managing a cluster of activities like ordering from multiple channels, inventories database analysis, order process including credit card verification, fulfillment systems, and return goods. This operation process also includes the entire order lifecycle starting from the placement of order and ensuring no missing of orders, or delay in the transition process.

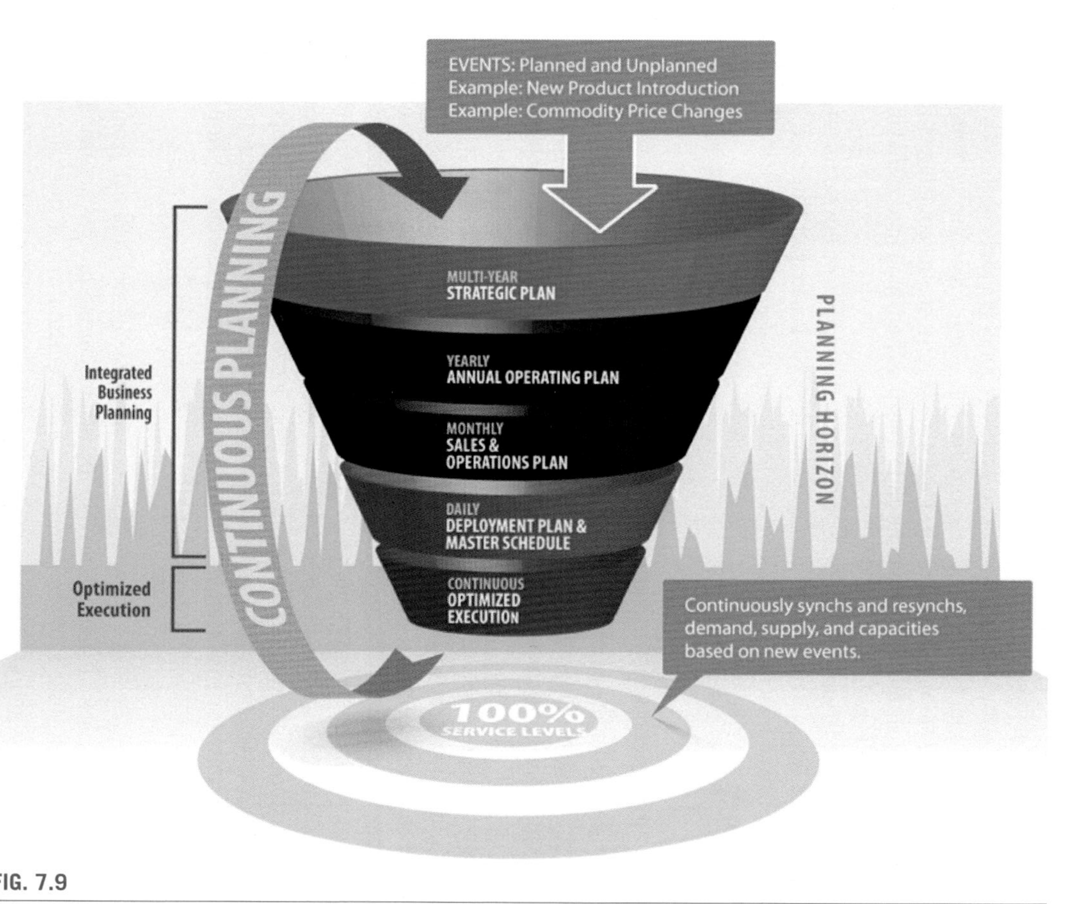

FIG. 7.9

The real-time value network, a model diagram.

A stock-keeping unit (SKU) is mainly developed on the basis of the items consumed by the surrounding population and their demand. The RTVN tool assists in how to operate the SKU of a specific locality. This practice helps in controlling and managing the inventories operation system. Order processing skill mainly depends on the dealings of a promoter with multiple parties. Sometimes a personal experience gives more effective results rather than by using some management tools. The best way of handling orders is to follow the cyclic manner of supplying goods, having adequate quantity in stock as inventory. The vendors should not feel harsh in receiving the goods in sufficient quality, and also timely. The supply chain mainly consists of manufacturers, distributors, and retail dealers. Generally, the manufacturers in collaboration with distributors manage the entire business transition in a cordial manner with each other (Fig. 7.10).

The entire order process operates with the starting point of the requirement of retail dealers. Manufacturers take timely action for a well-managed order and supply chain, after having thorough checking of credit, and customers need a base logistic system, even with the involvement of a third-party system.

7.7.1 Inventories procurement and utilization

Inventories are having economic value, and are kept in the warehouse or storehouses to meet the emergency requirement for the manufacturing process in order to maintain continuous harmonic manufacturing processes. In general, the inventories include consumable spares, or work in progress and finished goods, and are placed in a nearby location of the main manufacturing units. At wholesale dealers or retail dealers, the stocks are maintained on the demand basis of the customers on an as and when requirement basis. Besides this, a company has to maintain buffer stock of inventory to face unusual changes in demand. This practice helps a promoter in case of a sudden problem in the manufacturing process or delay in transition from the site of manufacture to wholesale dealers. Often, a wholesale dealer prefers to keep excess inventories by buying in bulk or orders more stock in advance of an impending price increase. Generally, the inventories are categorized into four groups to bring harmony in business management. These include: (i) raw material, (ii) in-process materials, (iii) finished goods, and (iv) MRO goods.

(i) Raw materials

For biopharmaceuticals manufacture, the inventories are different from other firms. The raw materials that are used in the manufacturing process are mainly microbial feedstock susceptible to pathogenic contamination, if not stored in proper conditions. In addition, the host cells used for biopharmaceutical product development are of animal, plant, and microbial origin. The methods and processes for host cells culture are highly sophisticated and need a digital control bioreactor of complex design.

These inventories may be commodities of biological origin that the firm or its subsidiary has produced or extracted. They also may be objects or elements that the firm has purchased from outside the organization. Even if the item is partially

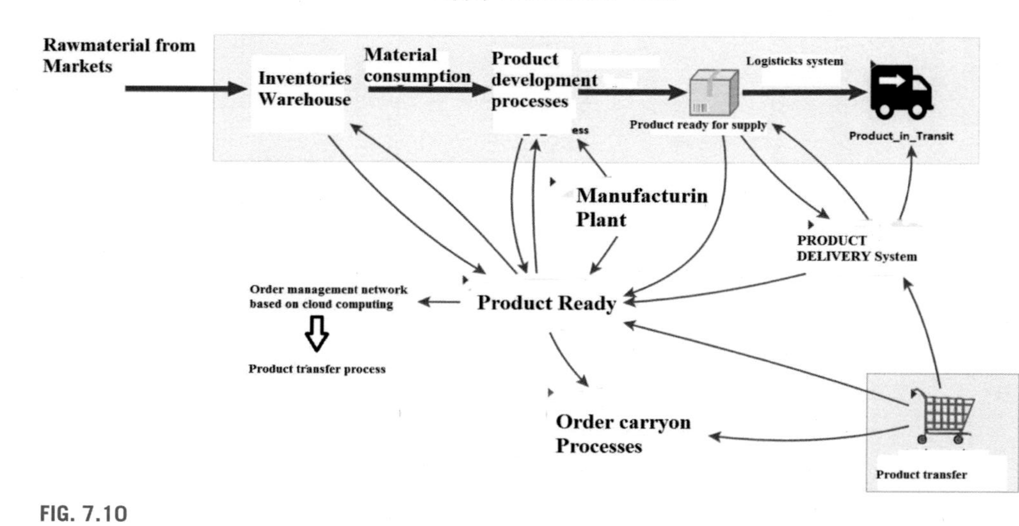

FIG. 7.10

A model on Supply Chain Management and order executive.

assembled or is considered a finished good to the supplier, the purchaser may classify it as a raw material if his or her firm had no input into its production.

The product structure tree clarifies the relationship existing within inventory items to understand the blueprint for the master production scheduled. For this purpose, the bill-of-materials file in a material requirement planning system (MRP) or a manufacturing resource planning (MRP II) plays a crucial role.

The best way of understanding the product structure tree is to understand the structure and function of a rolling cart. A rolling cart has a top made of steel or an ordinary iron sheet to carry goods. The structural stability is supported by a frame with four steel bars (as legs) with wheels to support the smooth movement of the rolling cart (Fig. 7.11).

In the manufacturing process in a biopharmaceutical industry, a final product (biologic) is produced through the assembly (formulation) of a variety of components which represent the raw materials. The raw material does not have any component. From the product structure tree point of view, a rolling cart's are steel, bars, wheels, ball bearing, axles, and caster frames.

(ii) Work-in-process (WIP)

In a biopharmaceutical industry, the biologically derived materials or the biomaterials under process for product development are known as work-in-process. In other words, any biological item originating from host cells, irrespective of plants, animals or microbes, but not from any feedstock, is considered to be work-in-process. WIP also includes the labor and overhead costs incurred for products that are at various stages of the production process. WIP is a component of the inventory asset account on the balance sheet, and these costs are transferred to the finished goods and eventually to cost of sale.

(iii) Finished goods

In a biopharmaceutical industry, the finished goods are defined as biological origin products manufactured from the industry under international acceptable regulatory acts and ready for packing and dispose to destination. These goods have been

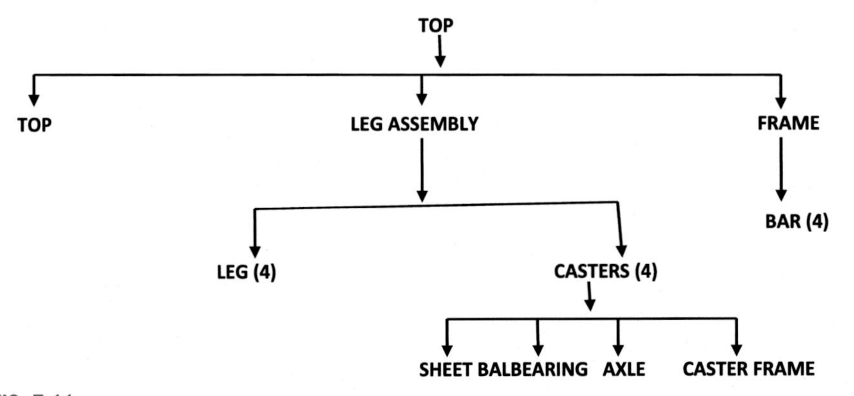

FIG. 7.11

Rolling carter with full assembly in graphic form.

inspected and have passed final inspection requirements so that they can be transferred out of work-in-process and into finished goods inventory. From this point, finished goods can be sold directly to their final user, sold to retailers, sold to wholesalers, sent to distribution centers, or held in anticipation of a customer order. Mostly, biopharmaceuticals are well managed to be in a temperature controlled condition for safe use. Inventories can be further classified according to the purpose they serve. These types include transit inventory, buffer inventory, anticipation inventory, decoupling inventory, cycle inventory, and MRO goods inventory. Some of these are also known by other names, such as speculative inventory, safety inventory, and seasonal inventory.

(iv) MRO goods

Generally, in the biopharmaceutical manufacturing process, MRO stands for maintenance, repair, and operation. MRO refers to any suppliers or goods that are used within the production process but are not part of the final product. MRO items include consumables (such as cleaning laboratory or office supplies), industrial equipment (such as compressors, pumps, valves), plant upkeep supplies (such as gaskets, lubricants, and repair tools) and computers fixtures, furniture, etc. In the biologics manufacturing process, first and foremost attention is to be given for maintenance of short life consumable items.

7.7.2 Other inventories

Transit inventory

Biopharmaceutical goods in the process of transferring from the packing room to the warehouse and subsequently loading in the logistics systems are known as transit inventory. It is also known as pipeline inventory. Merchandise shipped by truck or rail can sometimes take days or even weeks to go from a regional warehouse to a retail facility. However, most of the biopharmaceuticals, due to their sensitive property, are sent to the destination through freight by air. This practice not only saves time but also secures the quality of products, as desired by the consumers. Generally, the top biopharmaceutical industries employ freight consolidators to pool their transit inventories coming from various locations into one shipping source in order to take advantage of economies of scale. Of course, this can greatly increase the transit time for these inventories, hence an increase in the size of the inventory in transit.

Buffer inventory

During certain crisis periods, a biopharmaceutical manufacturing unit is not able to meet the demand of the manufacturing process. The raw materials used for the fermentation and culture of host cells are to be kept in sufficient as buffer inventory to meet the demand of the product manufacturing process and to maintain a sustainable supply chain. The buffer inventory acts as a cushion to meet the demand of the manufacturing process at the time of crisis. The sustainability of a firm's consumer service depends on the storage of sufficient buffer inventory.

Anticipation inventory

Sometimes biopharmaceutical industries in anticipation of a rise in the price of feedstock proceed to purchase sufficient feedstock to face the crisis such as a price increase, a seasonal increase in demand, or even an impending labor strike. Mostly, the retailers try to keep in advance inventory in anticipation of facing an unusual increase in price. When the demand is low, the manufacturers have the practice of keeping sufficient anticipation inventory for future demand without delaying in production time.

Decoupling inventory

The manufacturing process in a biopharmaceutical industry mainly depends on the upstream process of host cells mass culture methods and the downstream process is involved in product isolation, purification, crystallization, and polishing. In order to reduce the cost and process time, the various stages of the upstream process and downstream process are linked in a continuous or partially continuous manner. Due to some accidentally mechanical failure at the bioreactor operation stage or in the downstream process, the production rate gets slowed down or stopped.

However, in practice a promoter wants to keep the continuity and consistency in the manufacturing process. The work-in-process could be manageable with the proper arrangement of a decoupling inventory that serves as a shock absorber, cushioning the system against production irregularities. As such, it "decouples" or disengages the plant's dependence upon the sequential requirements of the system (i.e., one machine feeds parts to the next machine).

The more inventories a firm carries as a decoupling inventory between the various stages in its manufacturing system (or even a distribution system), the less coordination is needed to keep the system running smoothly. Naturally, logic would dictate that an infinite amount of decoupling inventory would not keep the system running in peak form. A balance can be reached that will allow the plant to run relatively smoothly without maintaining an absurd level of inventory. The cost of efficiency must be weighed against the cost of carrying excess inventory so that there is an optimum balance between inventory level and coordination within the system.

Cycle inventory

An inventory cycle is the fluctuation of GDP caused by the accumulation and selling of stocks or inventory. If production is greater than demand, GDP will rise but companies will also accumulate unsold stock. In supply chain management, the promoter of a manufacturing unit tries to balance inventory holding or carrying costs with the costs incurred from ordering or setting up machinery. This practice is known as economic order quantity (EOQ). When large quantities are ordered or produced, inventory holding costs are increased but ordering/setup costs decrease. Conversely, when lot sizes decrease, inventory holding/carrying costs decrease, but the cost of ordering/setup increases since more orders/setups are required to meet demand. When the two costs are equal (holding/carrying costs and ordering/setup costs), the total cost (the sum of the two costs) is minimized. Cycle inventories, sometimes called lot-size inventories, result from this process. Usually, excess material is ordered and consequently held in inventory in an effort to reach this minimization point. Hence, cycle inventory results from ordering in batches or lot sizes rather than ordering material strictly as needed.

Concept and significance of inventory management

Inventory management is the most important aspect of a biopharmaceutical industry. Due to the high-temperature sensitivity nature of biologics, the biopharmaceutical inventory in terms of product is most critical. In a biopharmaceutical industry, all functions are interlinked and connected to each other and are often overlapping. Some key aspects like supply chain management, logistics, and inventory form the backbone of the business delivery function. Therefore, these functions are extremely important to marketing managers as well as finance controllers.

Inventory management is a very important function that determines the potential of the supply chain as well as impacts the financial condition of the balance sheet. Every organization constantly strives to maintain optimum inventory to be able to meet its requirements and avoid over or under inventory that can impact the financial figures. Inventory is always dynamic. Inventory management requires constant and careful evaluation of external and internal factors and control through planning and review. Most of the organizations have a separate department or job function called inventory planners who continuously monitor, control, and review inventory and interface with the production, procurement, and finance departments. Every organization that is engaged in production, sale, or trading of Products holds inventory in one or the other form. While production and manufacturing organizations hold raw material inventories, finished goods and spare parts inventories, trading companies might hold only finished goods inventories depending on the business model.

7.7.3 Warehouse management

Warehouse management is the process, control, and optimization of warehouse operations from the entry of inventory into a warehouse. The warehouse manager supervises all activities of that facility, which often involves overseeing team performance, expending the receiving and shipping of goods, ensuring efficient and organization storage. Due to the perishable nature of biopharmaceuticals at elevated temperatures, a highly sophisticated and temperature regulated warehouse is needed for a temporary period of storing till a proper logistics system is ready for carrying the finished goods to end users, through wholesale dealers. Warehouse management is a key practice in delivering biopharmaceutical goods in safe condition. For biopharmaceutical products, the warehouse is fully air-conditioned with the provision of a digital controlling system to keep the products on the basis of needed temperature. Some bio-products like blood samples, fraction of animal and plant tissues, or therapeutic proteins need below the freezing point temperature for safe storing. The biopharmaceutical warehouse holds lifesaving medicines. A small variation in temperature may spoil the quality of biopharmaceutical products. That is why large companies cannot operate without state-of-the-art warehouse management solutions.

Temperature control warehouse for biopharmaceuticals

Warehouse for pharmaceuticals

FIG. 7.12

Types of warehouse for storage of biopharmaceuticals and pharmaceuticals.

The biopharmaceuticals distribution process is a cycle of complex actions that require the highest security standards and constant information sharing. Keeping full control over the system and having insight into even the smallest components determines the quality and efficiency of warehouse management (Fig. 7.12).

It ensures the constant availability and flow of essential quality health commodities, in appropriate quantities, in a timely and cost-efficient manner, through the supply chain system. Key warehousing functions include receiving and storing stock, inventory management, and distribution management. This brief focuses primarily on warehousing, even though warehousing and distribution are highly interrelated, and the same entity is often responsible for both functions. While storing stock is a key function of warehousing, the need for large warehouses and large holdings of stock may reflect inefficiency in the supply chain. In an ideal supply chain, large warehouses storing large volumes of products are unnecessary because products enter and exit the warehouse quickly and efficiently on their way to the service delivery point. The task before all supply chain practitioners is to determine how much storage space is truly necessary if operations are as efficient as possible.

Warehousing management in the biopharmaceutical industry and biotechnology industry has its own set of unique rules that have to be adhered to, in order to stock, pick, and ship product in a way that conforms to the many regulations that the market sector demands.

Storage areas should be designed or adapted to ensure good storage conditions. In particular, they should be clean and dry and maintained within acceptable temperature limits. Where special storage conditions are required on the label (e.g., temperature, relative humidity), these should be provided, checked, monitored, and recorded. Materials and pharmaceutical products should be stored off the floor and suitably spaced to permit cleaning and inspection. Pallets should be kept in a good state of cleanliness and repair (Fig. 7.13).

Storage areas should be clean and free from accumulated waste and vermin. A written sanitation program should be available, indicating the frequency of cleaning and the methods to be used to clean the premises and storage areas. There should also be a written program for pest control. The pest control agents used should be safe,

FIG. 7.13

Picking and packing at Cardinal Health's specialty warehouse.

Credit: Cardinal health.

and there should be no risk of contamination of the materials and pharmaceutical products. There should be appropriate procedures for the cleanup of any spillage to ensure complete removal of any risk of contamination.

Materials and pharmaceutical products should be stored in conditions which assure that their quality is maintained, and stock should be appropriately rotated. The "first expired/first out" (FEFO) principle should be followed. Recorded temperature monitoring data should be available for review. The equipment used for monitoring should be checked at suitable predetermined intervals and the results of such checks should be recorded and retained. All monitoring records should be kept for at least the shelf life of the stored material or product plus 1 year, or as required by national legislation. Temperature mapping should show uniformity of the temperature across the storage facility. It is recommended that temperature monitors be located in areas that are most likely to show fluctuations.

In recognition of the patient safety case, and the effect on reducing errors, it has been a regulatory requirement in the United States since 2006 that all pharmaceutical products sold to hospitals must now bear a bar code on the smallest unit of use, i.e., the size dispensed to the patient. Bar coding of medicines to the single unit also assists in the management of medicine alerts and recalls. The European Association of Hospital Pharmacists (EAHP) therefore calls on decision makers, politicians, and national administrations to implement bar coding of medicines to the single dose administered in hospitals as per national and European regulations. While storing biopharmaceutical and microbially derived products in a temperature controlled warehouse, the package should be well serialized with a proper bar code.

7.7.4 Transportation vs supply chain management

Transportation refers to the movement of product from the site of the manufacturing unit to the wholesale dealers from multiple locations (Fig. 7.14).

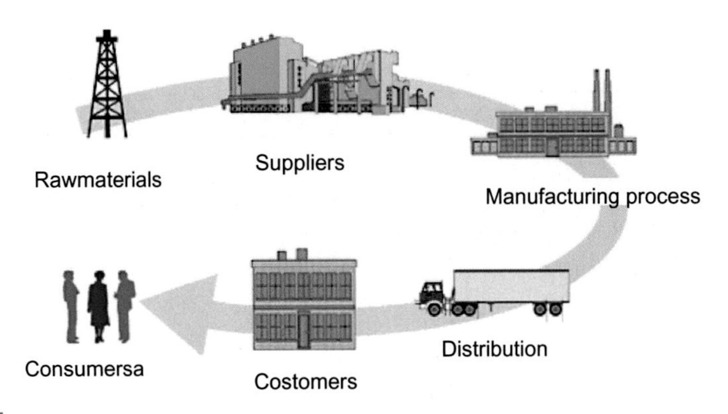

FIG. 7.14

Transportation supply chain Management.

Transportation is a part of supply chain management which includes procurement of goods from the site of manufacture, storing in proper condition before its distribution with the help of suitable logistics systems.

Many problems in the transportation supply chain can be addressed through the availability of analytics provided through a transportation management system. Needing such insights to allow companies to make smarter business decisions is especially true when supply chains become larger and begin operating on a larger scale. The recent advancements in IT technology also help promise better integration between the physical product movement and visibility. One good example is the surge of interconnected devices to connect pallets, trailers, and container systems in order to provide greater visibility. Of course, proper implementation is essential in order for these great technologies to succeed.

Generally, the biopharmaceutical industry has its own well-trained team to operate the supply chain for the safe delivery of goods on time. A well-executed logistic management system with digitally controlled tracing devices always leads to the greatest supply chain visibility. When transportation systems feed into a predictive analytics scheme, performance will be improved across the board. In fact, once the inventory is loaded into a channel, it is the predictive analytics responsibility to plan for efficient transportation. Most businesses would identify transportation supply chain visibility as a primary goal; however, it is sometimes used a little more than a marketing term that gets used to simply impress those who have invested in the company. This is especially true in larger enterprises. That is because many enterprises have never really figured out ways to implement real transportation and supply chain visibility. One of the reasons that transportation supply chain visibility does not get the attention it deserves is because as a process, supply chain visibility requires true system integration operating between many elements. Some of these elements have different master data that must be used. This data must be not only present, but also running in harmony over all systems. These can include warehouse management systems, multiple ERPs, ordering systems, and transportation management systems.

When these sources are made up across much geography, special attention is required in order to keep them all cohesively glued together. The glue is getting the materials from starting point to destination. The goal is to get this done efficiently and cost-effectively. Without transportation supply chain visibility, there will be time delays, expenses, and even backlogs. These events can throw off production schedules, even creating idle labor or eventually lost sales. So as the costs add up, you can see the importance of focusing on understanding transportation's role in the supply chain. Many problems in the transportation supply chain can be addressed through the availability of analytics provided through a transportation management system. The need for such insights to allow companies to make smarter business decisions is especially true when supply chains become larger and begin operating on a larger scale. The recent advancements in technology also help promise better integration between physical product movement and visibility. One good example is the surge of interconnected devices to connect pallets, trailers, and container systems in order to provide greater visibility. Of course, proper implementation is essential in order for these great technologies to succeed. Companies of all sizes must approach the transportation supply chain by implementing more harmonious systems in order to achieve greater visibility and a lower occurrence of supply chain errors. In the end, this will result in lower total costs for the organization, even beyond transportation costs. A well-executed transportation management system always leads to the greatest supply chain visibility. When transportation systems feed into a predictive analytics scheme, performance will be improved across the board. In fact, once the inventory is loaded into a channel, it is the predictive analytics responsibility to plan for efficient transportation.

7.7.5 Packing processes

In order to deliver a quality product, a planned packing system is an unavoidable factor. Due to the temperature sensitive nature of biopharmaceutical products, specially engineered packing is essential to protect the products from variable temperature, while in logistics systems. New innovations in raw materials packaging technologies can directly impact this process, streamlining operations, mitigation risks, and contributing to operational excellence (OpEx).

In the biopharmaceutical industry, two types of packing systems are in use: Traditional steel drum with plastic lining for large volume raw materials transport from the place of procuring to the site of manufacturing process are used, and smaller cardboard boxes with plastic liners having less than 50 kg are used for lesser volume biomaterials. Both bulk packaging systems are a part of the standard practices that most raw materials suppliers have established for their supply chain systems. The raw materials used in host cell cultivation include organic and inorganic salts, buffers, cell-cultured materials, mainly of biological origin, and are stored in a temperature controlled warehouse before use.

Containers and administration devices are integral to the safety and efficacy of biopharmaceutical therapies and must be key considerations for all new drugs coming on the market. By partnering with packaging manufacturers early in development, biopharmaceutical manufacturers can increase efficiencies in their production processes while gaining expert advice and counsel throughout a drug's lifecycle. Such an alignment can also improve product quality, enhance regulatory compliance, and contribute to the overall state of control by preserving biopharmaceutical stability, ensuring sterilization, and customizing delivery systems for market differentiation.

In recent years, an increasing number of biotherapeutics have received regulatory approval. This drives a need for continuous improvement and creates new requirements for primary packaging components. Enhanced regulations and new guidance on current good manufacturing practices (cGMP) are also driving drug and packaging manufacturers to improve production techniques. Biopharmaceuticals have unique packaging requirements for sophisticated container closure and drug delivery systems. These products are often more sensitive to interactions with vial and syringe components, which can induce protein aggregation or denaturization. Therefore, qualification and control of materials that contact biotherapeutics are critical to development and manufacturing. For highly sensitive biotech drugs, a packaging solution needs to offer protection from contamination and ensure consistency in every aspect that could affect the fill-finish process or long-term drug stability. Reducing variation in drug packaging can deliver measurable economic value by minimizing the risk of leachable in the drug solution. Several important concerns about biopharmaceutical packaging should be taken into consideration: diminishing and controlling visible and subvisible particles, ensuring consistency of component extractable, and maintaining consistent dimensions of packaging components throughout production to ensure manufacturing efficiencies and container closure integrity. Packaging components come with certified extractable profiles, which provide proof of each component's chemical consistency at a new level of control. Packaging manufacturers can also identify and control sources of potentially contaminating particles through inspection programs and continuous improvement efforts. All these factors combined can reduce or eliminate the expense and exposure to regulatory risk associated with component preparation, sterilization, and end-of-line rejections. The industry's leading suppliers can help guide drug product manufacturers through such issues and their potential implications.

The primary success in biopharmaceutical business is the design and development of an efficient method for the drug delivery process. In the recent past, the self-injection devices market has gained popularity because of the easy way of delivering liquid biologic drug through the self-injection method. This sort of practice makes a patient free from dependency on the doctor. So autoinjections represent an easy and convenient method which has been gaining popularity among the patients. Especially, patient suffering from chronic diseases like rheumatoid arthritis and multiple sclerosis feel easy for accepting such devices for regular drug delivery purpose without much dependence on the doctor's assistance. So drug delivery devices are the primary cause of success in marketing biologic drugs. So it is essential for a biopharmaceutical company to develop a customized delivery system (auto injector) before launching a liquid biologic in the market [10, 11].

7.8 Flow of manufacturing costs

The flow manufacturing cost is defined as the changeable cost of products starting from the procurement of raw materials incurred for manufacturing to the finished products. It is generally the practice of a biopharmaceutical company to assess the overall cost of the feedstock by summing up the cost of feedstock moves to work-in-process inventory. The total cost of feedstock is calculated on the basis of expenses incurred for the movement of raw material within the manufacturing preview. In the course of process flow, the costs are computerized at the beginning, and additional costs procured in the value chain process are added to calculate the overall cost of the product while preparing the final balance sheet (Fig. 7.15).

The overall manufacturing cost is calculated on the basis of inventories associated with the supply chain operation system. The major cost of production is dependent on the cost of inventory consisting of finished products for sale, raw material under process for producing the finished product. The calculation of profit is mainly dependent on sales revenue as well as all the cost involved in procuring raw material or live stock. The finalization of sales with a third party is decided by the competitive market scenario of the relevant product available from different sources.

In flow of manufacturing cost various terminology are being used to define the flow of manufacturing cost at different steps of goods transition from the site of production to the main store or warehouse and subsequently to the wholesale dealer. The two most common terminologies, i.e., FIFO (first in first out) and LIFO (last in first out), are in use in the main goods transition. Generally, the immediately manufactured goods are recorded as inventory. The costs of inventory items immediately after manufacture are changed to expense when the same are sold last. FIFO, on the other hand, assumes that the first items put on the shelf are the first items sold, so the old stocks are sold first. This system is generally used by the biopharmaceutical manufacturers where the inventories are perishable or prone to long storing practice. It is the mainly preferred accounting method when the cost of product is rising in the market. The cost flow mainly depends on the dynamic nature of the manufacturing process, work in progress, finished goods inventory, and quantitative procurement of raw materials or feedstocks for the manufacturing process. LIFO and FIFO each have a different effect on a manufacturer's financial condition.

A biopharmaceutical company, like any other manufacturer, uses the FIFO system to trace its inventory by reporting the LIFO method while preparing the final balance sheet. By this, the gap between FIFO and LIFO can be bridged by using LIFO as reserve. In case of price rising, a credit balance in the LIFO reserve is created, causing a reduction in inventory cost. Mostly, the manufacturers or business people use FIFO, or Standard Cost Methods, for internal use, and the FIFO method for external reporting, as the case tax preparation. By reporting in the LIFO method, this helps in reducing the tax at the time of price rising. Storing of goods in anticipation of demand and their marketing at the proper time are dependent on the efficacy of the supply chain management system. In the FIFO system, the manufacturer tries to clear the maximum amount of goods, and the products left without sale are mixed with

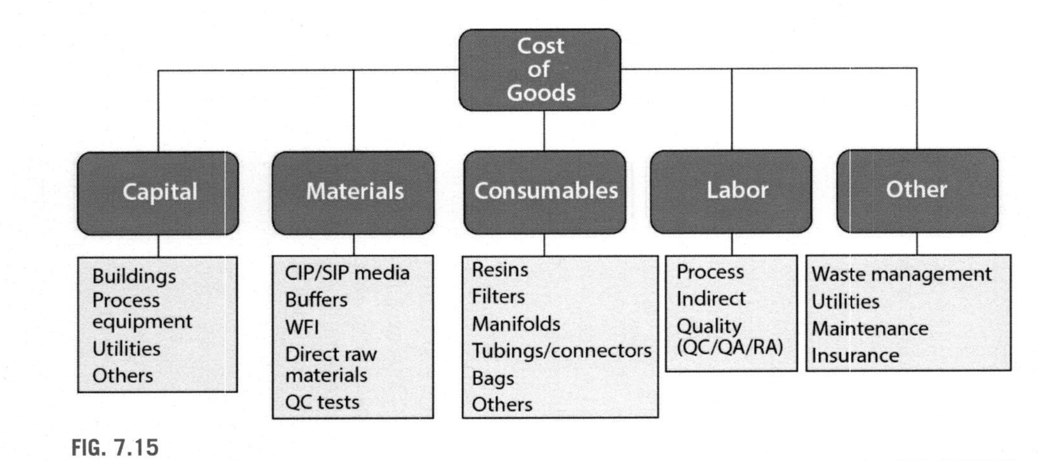

FIG. 7.15

Cost of goods categories for a biopharmaceutical manufacturing process.

the subsequent lot assuming that its quality matches with the existing leftover batch having a sufficient expiry date. The LIFO liquidation can be computerized in different ways. Generally, a manufacturer follows the LIFO liquidation by using the last in first out (LIFO) of inventory consisting, and then liquidation its older LIFO inventory. In case the current sale is higher than the manufacturing cost (including the cost of raw materials), the promoter should liquidate the stock inventory. This practice will be helpful in gaining financial profit in net operating income. Biopharmaceutical industries generally follow the LIFO method for immediate disposal of products in the market. This method is followed when the period of rising prices or period of inflation, when the cost to purchase inventory increases over time. The tax burden of a pharmaceutical company is reduced when the higher priced inventory matches the current revenue, increasing the cost of goods sold (COGS) and decreasing earning before (EBT). The LIFO method is used for asset management and valuation for asset management while acquiring, selling, or disposing of the inventory goods. If an asset is sold for less than it cost, then the difference is considered a capital gain. Using the LIFO method to evaluate and manage inventory can be tax advantageous, but it may also increase the tax liability.

7.9 Traceability and serialization techniques

Top multinational biopharmaceutical, pharmaceuticals, and biotechnology industries follow traceability and serialization techniques to meet the regulatory requirements, quality, and location tracing of goods for recall purposes [12]. For this purpose, the role of the community is of utmost significance to share timely information. This helps a company take quick action with the help of law enforcement at the national, state, and local levels. This method of regulation of goods enabled the US pharmaceutical industry to achieve over 50× reduction in pharmaceutical theft within a span of 4 years. Once upon a time, Eli Lily had the bad experience of missing drugs from the warehouse worth about $76 m [13]. The primary objective of traceability is typically not theft control, but to control quality and enable recall capability. The packed biopharmaceutical products should carry a tracking microchip to understand the condition of goods instantly.

7.10 Expiration date determination

In the biopharmaceutical industry, the life span of a product is most important. Each packing should carry a sticker mentioning the date of expiration. An expiration date means the maximum life of a biological origin product, after which it is strictly not advised for further use, and should be destroyed, without creating any environmental problem. Arbitrary expiration dates are also commonly applied by companies to product coupons, promotional offers, and credit cards. In these contexts, the expiration date is chosen for business reasons or to provide some security function rather than any product safety concern. Expiration date is often abbreviated as EXP. The legal definition and usage of the term expiration date varies between country and nature of the products. The expiration date for each biological origin product should be computed from the date of the initiation of the potency test. Prior to license, the stability of each fraction should be determined by methods acceptable to the animal and plant health inspection service.

7.11 Modernization of supply chain operation

A steady increase in demand of biopharmaceuticals, nutraceuticals, food additives, and biosimilars has perused the multinational groups of the efficacy of increasing manufacture process and supply chain management, in order to meet the demand of end users without much financial burden. To bring success in this practice, information technology (IT) and supply chain management are unavoidable factors in bringing sustainability in the manufacturing process [14]. Factually, IT technology creates a strong link with the supply chain and helps in keeping close touch with consumers. This practice is ultimately helpful in the manufacturing process and quality control. Thus, the alignment of IT and SCM can increase the production efficiency and lead a company toward profitability [15]. Integration of SC and IT technology brings immense benefit in product development, procurement, manufacturing, physical distribution, customer relationship management, and performance measurement [16]. This sort of practice provides information to a promoter about the status of the existing link between technology and people's understanding [17].

Facilitation of the supplier-customer relationship is maintained by well-managed SCM. The competitive attitude of a company develops a defensible position over its competitors. Since long back, the promoters of different company have been experienced the benefit of empirical research on cost, quality, delivery, and flexibility in business management [18]. The advantage of the present supply chain system over the traditional supply chain was multifolds [19]. The application of IT technology in SCM has quickened the business activities, improved decision-making, and productivity. This is accomplished through the exploitation of IT for internal and external integration of business processes.

By liking IT technology, a company can integrate internal business function and keep track of the activities of the company well in time. This can help the promoter to fulfill the requirements of the customer and catalyze the quality control process. [20]. SCM systems with the help of IT help logistics management, transportation management, strategic planning, warehousing, inventory, manufacturing, supplier management, and customer management [21]. Enterprise Resource Planning (ERP) systems are included as part of the broader SCM software. ERP systems are employed to integrate business processes by organizing, codifying, and standardizing business processes and data [22]. Functionally, ERP helps in systematic way of data management and transferring data within the internal system. ERP data mines can be used to forecast production and make decisions.

Another key factor in the supply chain process is customer relationship management (CRM), which is the management skill of relationships between the organization and its customers. Integration of ERP with CRM helps in the internal integration of business. Many ERP vendors bind with the ERP-CRM integration package to have a better CRM system. Moreover, the establishment of ERP and CRM systems used to be a concern with enterprises involved in embarking on e-business [23] to share critical information on demand prediction, actual orders, and inventory level in an efficient manner [24], while protecting each company's proprietary data. The adoption of inter organizational information systems (IOS) was for external integration [25, 26].

There are some technologies that overall are considered of high relevance for future supply chains, such as cloud solutions and mobile devices. Even though security concerns are under consideration, cloud computing has become an integral part of today's supply chain management solutions portfolio and remains of interest. The modern IT base supply chain software tools make the company easier and convenience to assess the investment at various stages of development, and make easier to keep touch with global trading. Mobile devices have facilitated the business transition process and acquainted the promoter on updated information related to the supply chain management process.

A good logistics-based supply chain management system can fulfill the requirements of the 21st century in order to transport and supply goods on time in a safe and secure manner.

7.12 Green logistics and supply chain

The advanced digital-based logistics systems management has tremendously lessened the burden of pollution. Emission of greenhouse effective gaseous pollutants has reduced and is expected to further reduce with the implementation of international understanding on greenhouse effect and global warming. Logistics systems development and operation keeping in view of global pollution is otherwise known as green logistics and supply chain (GLSC). It also includes green storage, green packaging, green circulation processing, green recovery, and other activities via advanced logistics technology. It aims to reduce emission of gaseous pollutants arising from logistics activity so as to realize a "win-win" consequence in logistics development and eco-environmental conservation. In this process, the traditional practice of supply chain is integrated with environmental conservation so that the waste generated during any supply chain can be well recycled in order to induce susceptibility in the supply chain management practice and bring profit to the manufacturer (Fig. 7.16).

Microbial processes activity for the production of biopharmaceutical, health-care product like nutraceuticals, food supplements, food additives, antioxidants, etc. may be a threat to the environment in terms of carbon monoxide emissions, discarded packaging materials, scrapped toxic materials, traffic congestion, and other forms of industrial pollution.

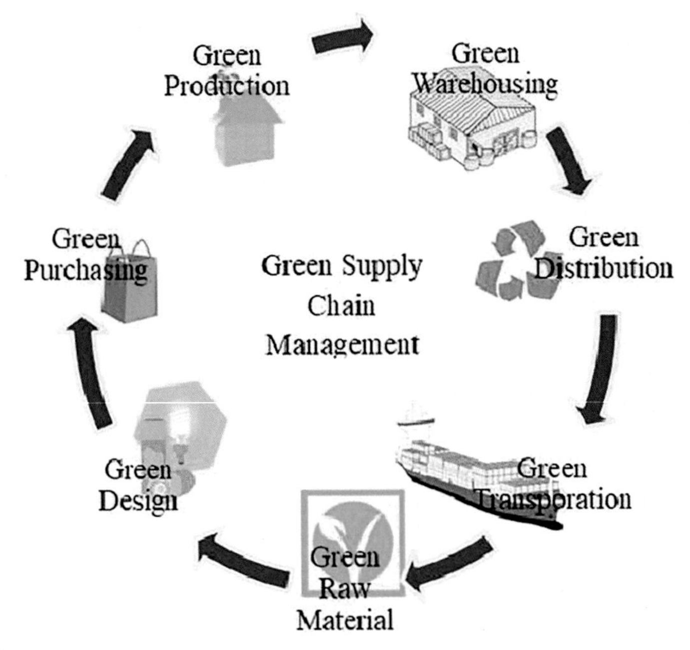

FIG. 7.16

Green supply chain management system.

Green logistics and supply chain (GLSC) impacts on the environment are minimum harm on the sustainability performance of the supply chain. During the beginning of the 1990s, competition among the big multinationals increased in order to impress the consumers from the aesthetic and social responsibility points of view on product purity maintenance, without bringing any issue on environmental cleanness [27]. In due course of time, GSCM has impressed and could convince the manufacturer of the real market demand by adapting this technique [28]. Having these practices in mind, firms develop environmental management strategies in response to the dynamic nature of environmental requirements and their impacts on supply chain operations. A fruitful supply chain management is an integrated process of supplying goods to the end user by means of a wide network consisting of all the parties involved (e.g., supplier, manufacturer, distributor, wholesaler, retailer, customer, etc.), in manufacturing a cGMP certified product, and its delivery with the assurance of traceability during transportation to the retail dealer or services to the ultimate customers both upstream and downstream.

7.13 Quality control

Quality control is the main backbone of success in developing and managing a biopharmaceutical manufacturing company. As the biopharmaceuticals are biological in origin, the products are susceptible to microbial contamination and elevated temperature, and quality control is unavoidable to keep the product safe till it reaches the end users. Due to an increase in quality awareness, it has been mandatory to incorporate a blueprint of quality control while establishing an industrial undertaking program. Proper quality control ensures the most effective utilization of available resources and reduction in cost of production.

The word quality control comprises two words, viz., quality and control. It would be appropriate to explain these two words separately to understand clearly the meaning of quality control. According to Dr. W.K. Spriegel, "*The quality of a product may be defined as the sum of a number of related characteristics such as shape, dimension, composition, strength, workmanship, adjustment, finish and color.*" In the words of John D. McIlellan, "Quality is the degree to which a product conforms to specifications and workmanship standards." It is clear from these definitions that quality refers to the various characteristics of a product and their excellence. Quality is a relative term and is never absolute depending upon the use of the product and circumstances under which it is used.

Objectives of quality control

The main target of quality control is to manufacture consumer acceptable product(s) and make it reachable to the end user with the guarantee of a quality product. Besides this, quality control also takes the responsibility of a flawless manufacturing process by suggesting improvements in the manufacturing process, if demanded by the users. Successful businesses inevitably place great emphasis on managing quality control by carefully planned steps taken to ensure that the products and services offered to their customers are consistent and reliable and truly meet their customers' needs. Multinational corporations have entire departments of highly trained specialists to design and implement their quality assurance programs. It is said that when the Japanese business that later became Sony Corporation was founded, the cofounder Mr. Ibuka established the company philosophy by stating that "*If it were possible to establish conditions where persons could become united with a firm spirit of teamwork and exercise to their hearts' desire their technological capacity, then such an organization could bring untold pleasure and untold benefits.*" For decades, Japanese businesses have pioneered management techniques intended to improve quality continuously. American and European companies have spent years catching up. In order to maintain or enhance the quality of the offerings, manufacturers use two techniques, quality control management and quality assurance. These two practices make sure that the end product or the service meets the quality requirements and standards defined for the product or the service.

7.13.1 Quality control management

Quality control (QC) and quality assurance (QA) are two separate entities in manufacturing a customer oriented product. The process of making sure that the stakeholders are adhering to the defined standards and procedures is called quality control. In quality control, a verification process takes place. Every organization that practices

QC needs to have a Quality Manual. The quality manual outlines the quality focus and the objectives in the organization. The quality manual gives the quality guidance to different departments and functions. Therefore, everyone in the organization needs to be aware of his or her responsibilities mentioned in the quality manual.

There are many methods followed by organizations to achieve and maintain the required level of quality. Some organizations believe in the concepts of Total Quality Management (TQM), and some others believe in internal and external standards. The standards usually define the processes and procedures for organizational activities and assist in maintaining the quality in every aspect of organizational functioning. When it comes to standards for quality, there are many. ISO (International Standards Organization) is one of the prominent bodies for defining quality standards for different industries. Therefore, many organizations try to adhere to the quality requirements of ISO. In addition to that, there are many other standards that are specific to various industries. As an example, SEI-CMMi is one such standard followed in the field of software development. Since standards have become a symbol for products and service quality, customers are now keen on buying their product or service from a certified manufacturer or service provider. Therefore, complying with standards such as ISO has become a necessity when it comes to attracting customers. The importance of properly established and managed quality control and quality assurance systems with their integral well-written SOPs and other quality documents for the achievement of the Company's business objectives cannot be ignored. They serve as a passport to success and maintain sustainability in business.

7.13.2 Quality assurance

Quality assurance is a broad practice used for assuring the quality of products or services. There are many differences between quality control and quality assurance. In quality assurance, constant effort is made to enhance the quality practices in the organization.

The quality assurance department must be autonomous in nature, and operate without any threat and pressure from a higher administrative body, unless there is a need-based demand of users as and when required. Quality assurance should be under the strict guidance of Regulatory Acts like GxPs: current Good Manufacturing Practice (cGMP), Good Laboratory Practice (GLP), Good Clinical Practice (GCP), etc., and local, national, regional, and international legal, ethical, and regulatory requirements. Therefore, continuous improvements are expected in quality functions in the company. For this, there is a dedicated quality assurance team commissioned. Sometimes, in larger organizations, a 'Process' team is also allocated for enhancing the processes and procedures in addition to the quality assurance team. The quality assurance team of the organization has many responsibilities. First and foremost, the responsibility is to define a process for achieving and improving quality.

Some organizations come up with their own process and others adopt standard processes such as ISO or CMMi. Processes such as CMMi allow the organizations to define their own internal processes and adhere by them. The quality assurance function of an organization uses a number of tools for enhancing the quality practices. These tools vary from simple techniques to sophisticated software systems. The quality assurance professionals also should go through formal industrial training and get certified. This is especially applicable for quality assurance functions in software development houses. The quality assurance teams of organizations constantly work to enhance the existing quality of products and services by optimizing the existing production processes and introducing new processes. Following are two important practices in quality assurance:

(1) *Stability test*

Biotechnological and biopharmaceutical products are extremely susceptible to environmental stress and strain. By the stability test, one can understand the change in quality of such a product in response to temperature, humidity, light like various physical factors while storing and transporting to the destination. The standard protocol for testing is well defined on the basis of effects of climatic conditions in the three regions of the EU, Japan, and the United States.

On the basis of global climatic data, one can easily calculate the instant climatic condition of a specific region. On average, the climatic condition of the world can be divided into four specific zones. This guidance addresses climatic zones I and II. The principle has been established that stability information generated in any one of the three regions of the EU, Japan, and the United States would be mutually acceptable to the other two regions, provided the information is consistent with this guidance and the labeling is in accord with national/regional requirements.

The FDA's guidance documents, including this guidance, do not establish legally enforceable responsibilities. Instead, guidance's described the Agency's current thinking on a topic and should be viewed only as recommendations, unless specific regulatory or statutory requirements are cited. The use of the word should in Agency guidance's means that something is suggested or recommended, but not required.

Stress testing can be helpful in understanding the degradation nature of biomolecules, which can ultimately be a guideline in studying the biological conservation of biotechnological and biological products during the distribution pattern to the consumer. In practice, stress testing is likely to be carried out on a single batch of the drug substance. The testing should include the effect of temperatures [in 10°C increments (e.g., 50°C, 60°C) above that for accelerated testing], humidity (e.g., 75% relative humidity or greater) where appropriate, oxidation, and photolysis on the drug substance. The testing should also evaluate the susceptibility of the drug substance to hydrolysis across a wide range of pH values when in solution or suspension. Photostability testing should be an integral part of stress testing. The standard conditions for photostability testing are described in ICH *Q1B* Photostability Testing of New Drug Substances, and Products analysis for the stability study should be carried out with a minimum of three batches. The capacity of each batch should not be less than the size of pilot level production followed by the same synthetic pathway as practices in turnkey level production. The overall quality of the batches of drug substance placed on formal stability studies should be representative of the quality of the material to be made on a production scale. The stability studies should be conducted on the drug substance packaged in a container closure system that is the same as or simulates the packaging proposed for storage and distribution. In the stability test, more attention should be given to the biological, which is susceptible to change during storage and transport in order to keep vigilance on the efficacy and quality of the biological product. The testing should cover, as appropriate, the physical, chemical, biological, and microbiological attributes. Validated stability-indicating analytical procedures should be applied. Whether and to what extent replication should be performed would depend on the results from validation studies. For long-term studies, the frequency of testing should be sufficient to establish the stability profile of the drug substance. For drug substances with a proposed retest period of at least 12 months, the frequency of testing at the long-term storage condition should normally be every 3 months over the first year, every 6 months over the second year, and annually thereafter through the proposed retest period. Well-defined protocols are available on storage of biological/biotechnological/ pharmaceutical products under variable physical conditions [29–36].

Generally, the biologic's potency depends on the conjugation of the active substance(s) to a second moiety or binding to an adjuvant. Dissociation of the active substance(s) can be examined under real-time/real temperature conditions (including conditions encountered during shipment).

The following product characteristics, although not specifically mandatory for biologics, can be applicable for better understanding:

(i) Visual appearance of the product like color, opacity for solutions/suspensions, texture, and dissolution time. Besides this, the presence of visible particulates in solutions or after the reconstitution of powders or lyophilized cakes, pH, and moisture level in the powder and lyophilized products can be considered as important factors while characterizing biologics.
(ii) Sterility testing or alternatives (e.g., container/closure integrity testing) should be performed in the beginning or at the end of the proposed shelf life.
(iii) Additives (e.g., stabilizers, preservatives) or excipients may degrade during the dating period of the medicinal product. So traceability of degraded additives is important while characterizing products. Studies that reaction or degradation of such materials adversely affect the quality of the medicinal product; these items may need to be monitored during the stability program.
(iv) The container/closure has the potential to adversely affect the product and should be carefully evaluated.

(2) *Existing legal basis for approval of biologics*

Mainly, two US statutes apply to the regulation of biological products, which include: (i) the Federal Food, Drug, and Cosmetic Act (21 USC § 301 et seq) (FFDCA), and (ii) the Public Health Services Act (42 USC § 262) (PHSA). The US regulations control the FDA for monitoring the safety and security of biologics. The FDA administers include: the FFDCA and PHSA (among other statues). The United States Federal Food, Drug, and Cosmetic Act applies to all drugs and medical devices, and PHSA applies to "biological products." Marketing approval under the FFDCA is by means of a New Drug Application (NDA) while approval under the PHSA is by means of a Biologics License Application (BLA). Both drugs and biologics are subject to Investigational New Drug Application (INDA) regulations. Preclinical research on new compounds is carried out in a laboratory, using a wide variety of techniques. Promising candidates are then studied in animals, and subsequently, various clinical studies in humans are carried out following strict guidelines: Phase I: A small number of healthy volunteers are given the compound to determine mainly that the drug is safe for human use, and Phase II: A small number of patients are given the medicine to assess its efficacy and safety and to ensure that there are no unacceptable side effects. Phase III: A large number of patients, usually thousands, take the medicine under supervision over a defined period of time, with the results used to establish efficacy.

If the results show the drug to be efficacious and safe, the data are presented to the FDA. The FDA reviews the data, and if the data is acceptable, a marketing authorization is issued. Alternatively, the FDA may request additional studies or reject the application. Following the grant of marketing authorization, the drug product is studied in large numbers of patients in hospitals and clinics to further assess its clinical effectiveness. This stage is called Phase IV or the postmarketing study. Safety Assessment of Marketed Medicines (SAMM) studies help identify any unforeseen side effects. In order to be marketed, a biologic requires only proper labeling and an approved BLA that indicates the product has been determined to be safe, pure, and potent and that the manufacturing facilities meet the requirements to ensure safety, purity, and potency. Though biologics have traditionally been subject to much more scrutiny in manufacturing than drugs, those differences are being eroded. Biologics have been approved under the FFDCA and PHSA; thus, both NDA and BLA applications have been submitted for biologics. The exceptions are glucagon and follistim, which were approved under § 505(b)(2), and insulin, which was approved under its own statute for a time. The default approval pathway for biologics now is a BLA, unless the product is a hormone, in which case §505(b) is used.

7.13.3 Quality control and supply chain integration

The past two decades have witnessed the positive impact of total quality management (TQM) on acquiring success in biotechnological and biopharmaceutical business [37, 38]. It plays a vital role in the development of management practices [39]. It has been noticed that TQM is immensely helpful in upgrading the business transition by fulfilling the complete requirement of customers [40]. In addition, TQM increases high competitive in business at national and global market [41] and above all as a source of enhancing organizational performance through continuous improvement in the organization's activities [42–44]. Henceforth, it is strongly recommended to incorporate quality principles and procedures in SCM. Thus, quality management is often related to a model of organizational change [45], the implementation of which largely relies on the organization's ability to adapt itself to these principles.

Quality is an important factor when it comes to any product or service. With the high market competition, quality has become the market differentiator for almost all products and services. So the practice of QC has been an integral part of the supply chain in the production and distribution of product at the consumer level. It helps in developing confidence in the customer on product acceptability. Without quality control, waste becomes prevalent beyond a tolerable amount. For biotechnology and biopharmaceutical industries, feedstock for bacterial growth and development is abundant. How to procure quality livestock depends on the quality control wing of the company. So the quality control team controls the quality of livestock and its proper storage. Genetically engineered microbes are in used to produce value-added biochemicals. Quality control helps in screening those genetically modified microbes from the possibility of possessing toxin compounds harmful for human use. The US Department of Transportation prescribes important rules for the transport of hazardous substances. Noncompliance can lead to penalties or fines, which makes quality control imperative. The more efficiently and effectively toxic materials are handled in the supply chain, the better it will be for all internal and external stakeholders.

References

[1] Anon. 3PL (Third party logistics): select the right fulfillment partner in a pandemic, www.shopify.in; 2020.
[2] Global Clinical Trial Supply and Logistics Market For Pharma 2017–2027 Visiongain Ltd - https://www.globenewswire.com/news-release/2017/05/10/981838/0/en/Global-Clinical-Trial-Supply-and-Logistics-Market-for-Pharma-2017-2027.html.
[3] Zekhnini K, Cherrafi A, Bouhaddou I, Benghabrit Y, Garza-Reyes JA. Supply chain management 4.0: a literature review and research framework. BIJ 2020;28(2):465–501. https://doi.org/10.1108/BIJ-04-2020-0156.
[4] Spekman RE, Carr R. Making the transition to collaborative buyer–seller relationships: an emerging framework. Ind Mark Manag 2006;35:10–9.
[5] Lamming RandHampson J. The environment as a supply chain management issue. Br J Manag 1996;7:S45–62.
[6] Saunders MJ. Strategic purchasing and supply chain management. London, UK: Pitman; 1997.
[7] Saunders MJ. Making strategic decisions and actions in purchasing and supply chain management. In: Proceedings of the 6th international IPSERA conference. Italy: Naples; 1997. p. 61–9.
[8] Croom S, et al. Supply chain management: an analytical framework for critical literature review. Eur J Purch Supply Manag 2000;6(1):67–83.
[9] Harland CM. Supply chain management, purchasing and supply management, logistics, vertical integration, materials management and supply chain dynamics. In: Slack N, editor. Blackwell encyclopedic dictionary of operations management. UK: Blackwell; 1996.
[10] US Food and Drug Administration. Current good manufacturing practice for finished pharmaceuticals. Code of Federal Regulations, Title 21, Part 211. Rockville, MD: US Food and Drug Administration; 2017. www.accessdata.fda.gov.
[11] 2017 Biopharma Cold Chain Sourcebook www.hexaresearch.com. Third party logistics market size - 3PL industry report, 2014, www.pharmaceuticalcommerce.com; 2018.

[12] McBeath B. How the pharmaceutical industry has dramatically reduced thefts. Chainlink Research www.chainlinkresearch.com; 2012.
[13] Ketchen DJ, et al. Best value supply chains: a key competitive weapon for the 21st century. Bus Horiz 2008;51:235–43.
[14] Kozlenkova IV, Hult G, Tomas M, Lund DJ, Mena JA, Kekec P. The role of Marketing channels in supply chain management. J Retail 2015;91(4):586–609.
[15] Chandra C, Grabis J. Supply chain configuration- concepts, solutions and applications. Springer, New York: Springer Science+Business Media; 2007.
[16] Olson LD. Supply chain information technology. In: Nahmias S, editor. The supply and operations management collection. New York: Business Expert Press; 2012.
[17] Shaik MN, Abdul-Kader W. Interorganizational information systems adoption in supply chains: a context specific framework. Int J Supply Chain Manag 2013;6(1):24–40.
[18] Nelson M. In: Papp R, editor. Sustainable competitive advantage from information technology: limitations of the value chain; 2001.
[19] Ravindran R, Warsing Jr D. Supply chain engineering: models and applications. CRC Press; 2017, ISBN:9781138077720.
[20] Hines T. Supply chain strategies: demand driven and customer focused. Taylor & Francis; 2014, ISBN:978-1-136-70396-6.
[21] Turek B. Information systems in supply chain integration and management; 2013. Retrieved from http://www.ehow.com/ info_8337099_information-supply-chain-integration-management.html.
[22] Norris G, et al. E-business and ERP: transforming the enterprise. Canada: John Wiley & Sons; 2000.
[23] Yanjing J. Integration of ERP and CRM in E-commerce environment. In: Proceedings of the international conference on management and service science; 2009. p. 1–9.
[24] Marakas MG. Decision support systems in the 21st century. Prentice Hall: Indiana University, NewJersey; 2003.
[25] Premkumar GP. Interorganization systems and supply chain management: an information processing perspective. Inf Syst Manag 2000;17(3):56–69.
[26] Grover V, Saeed KA. The impact of product, market, and relationship characteristics on interorganizational system integration in manufacturer supplier dyads. J Manag Inf Syst 2007;23(4):85–216.
[27] Diabat A, Govindan K. An analysis of the drivers affecting the implementation of green supply chain management. Resour Conserv Recycl 2011;55(6):659–67.
[28] Luthra S, et al. Green supply chain management. Implementation and performance—a literature review and some issues. J Adv Manag Res 2014;11(1):20–46.
[29] ICH Q1B. Photostability testing of new drug substances and products. In: International conference on horminisation of technical requirements for registration of pharmaceuticals for human use. Center for Biologics Evaluation and Research (CBER), ICH; 1996. www.ich.org/fileadmin/Public_Web_Site/ICH_Products/.../Q1B/.../Q1B_Guideline.pd.
[30] ICH Q1C. Stability testing for new dosage forms. Center for Biologics Evaluation and Research (CBER); 2006. www.ich.org/fileadmin/Public_Web.../ICH.../Guidelines/.../Q1C/.../Q1C.
[31] ICH Q3A. Impurities in new drug substances. Center for Biologics Evaluation and Research (CBER); 2006. www.ich.org/fileadmin/Public_Web_Site/ICH.../Q3A.../Q3A_R2.
[32] ICH Q3B. Impurities in new drug products. Center for Biologics Evaluation and Research (CBER); 2006. www.pharma.gally.ch/ich/q3b028295en.pdf.

[33] ICH Q5C. Quality of biotechnological products: stability testing of biotechnological/biological. Center for Biologics Evaluation and Research (CBER), Products; 1995. www.ich.org/products/.../quality/quality.../stability-testing-of-biotechnologicalbiologi.
[34] ICH Q6A Specifications. Test procedures and acceptance criteria for new drug substances and new drug products: chemical substances. Center for Biologics Evaluation and Research (CBER); 2000.
[35] Anon, www.ich.org/fileadmin/Public_Web.../ICH_Products/Guidelines/.../Q6A/.../Q6Astep4.
[36] ICH Q6B Specifications. Test procedures and acceptance criteria for new drug substances and new drug products: biotechnological/biological products. Center for Biologics Evaluation and Research (CBER); 2005. www.ich.org/fileadmin/Public_Web_Site/ICH_Products/.../Q6B/.../Q6B_Guideline.pd.
[37] Arumugam V, et al. Self-assessment of TQM practices: a case analysis. TQM J 2009;21:46–58.
[38] Mohanty RP, Behera AK. TQM in the service sector. Work Study 1996;45:13–7.
[39] Hoang DT, et al. The impact of total quality management on innovation: findings from a developing country. Int J Qual Reliab Manag 2006;23:1092–117.
[40] Oakland JS. Total quality management. 2nd Ed. Oxford: Butterworth-Heinemann; 1993.
[41] Terziovski M. Quality management practices and their relationship with customer satisfaction and productivity improvement. Manag Res News 2006;29:414–24.
[42] Claver-Cortés E, Pereira-Moliner J, Tarí JJ, Molina-Azorín JF, et al. TQM, managerial factors and performance in the Spanish hotel industry. Ind Manag Data Syst 2008;108:228–44.
[43] Teh PL. Does total quality management reduce employees' role conflict? Ind Manag Data Syst 2009;109:1118–36.
[44] Blackiston GH. A barometer of trends in quality management. Natl Prod Rev 1996;16:15–23.
[45] Boronat P, Canard F. Management par la qualite: A totale et changementorganisationnel. Paris, Economica: Les NouvellesFormsOrganisationnelles; 1995.

CHAPTER

Primary health-care goal and principles 8

8.1 Introduction

Primary health care (PHC) refers to a broad range of health services provided by medical professionals in the community. This means universal health care is accessible to all individuals and families in a community.

General health-care practitioners, nurses, pharmacists, and allied health-care providers are exclusive components of the primary health-care team. Basically, the PHC service is the process and practice of immediate health services, including diagnosis and treatment of a health condition, support in managing long-term health care, including chronic conditions such as diabetes. PHC also includes regular health checks, health advice when an unhealthy person seeks support for ongoing care (Fig. 8.1). In India, the government has fixed specific norms for primary health center, based on of community structure and population (Table 8.1).

8.2 Conceptual development of PHC

India is the first country to implement primary health services before the Declaration of Alma-Ata. The basic motto for adapting primary health service is to serve the people to maintain health without spending money from the pocket. On the basis of the Health Survey and Development Committee Report of 1946, the Indian Government implemented primary health service in rural community [1].

Many projects such as the most acclaimed The National family planning programme (launched in 1952) and the policy of one community health worker per 1000 people in the 1970s have been implemented for bolstering its health-care scenario. In 2005, the UPA Government launched the Rural Health Mission (NRHM), as a move to improve access to quality health care, especially for poor rural women and children. These entire health-care-related projects resulted in a remarkable decline of maternal mortality ratio (MMR) by 77% from 556 per 100,000 live births in 1990 to 130 per 100,000 live births in 2016.

Since the late 1960s and early 1970s, the concept on primary healthcare has emerged in the United States for the first time, fighting against malaria at the community level. The government initiated a health-care project at the community level. WHO staff members conducted a survey and studied the experiences of medical auxiliaries in developing countries and argued that "a strict health sectoral approach

Healthcare Strategies and Planning for Social Inclusion and Development. https://doi.org/10.1016/B978-0-323-90446-9.00008-3

FIG. 8.1

Primary health-care center and services to community population (view from an Indian village).

Table 8.1 Primary health-care structure and their structural norms (rural health infrastructure in India).

Center	For plain area (ft^2)	For hilly tribal area (ft^2)
Subcenter	5000	3000
Primary health care	30,000	20,000
Community healthcare	120,000	80,000

is ineffective" [2]. In 1874, Canadian Lalond report de-emphasized the importance attributed to the number of medical institutions and proposed determinants for health biology, health services, environment, and lifestyle [3].

In the late 1960s, Christian Medical Commission started planning on PHC. This organization, with the help of the World Council of Churches, and the Lutheran World Federation stated mission on PHC to emphasize the training of grassroot village workers equipped with very essential drugs and simple methods. In 1970, it published a Journal "*Contact*" based on "primary health care." Subsequently in 1974 WHO extended support on further popularization of this journal at the global level [4].

In 1948, World Health Organization (WHO) initiated the agenda for primary health care, and later it was highlighted through Alma-Ata International Conference on Primary Health care [5, 6]. Basically, primary health care is for the community/whole society for a wide range of health services, including health promotion; diseases prevention; treatment and rehabilitation; and palliative care [7]. The entire process for primary health-care services is based on population dynamics and the health-care system in order to integrate personal health care, public health function, and hospital management process. The provision of primary health-care services is only a part of the broad spectrum of primary health-care concepts to address the determinants of health; the implementation of a primary health-care concept must be accompanied by multispectral actions and the empowerment of the population [8].

Since the last four decades, PHC Provision has moved from simple planning to action. Alma-Ata Declaration, is still crucial in current global health like COVID-19 pandemic, especially for developing countries. To achieve this, initially, the United Nations (UN) announced eight Millennium Development Goals (MDGs) by 2015 [4]. But looking at the progress, the UN announced 17 Sustainable Development Goals (SDGs), with the strong hope of achieving the same by the end of 2030 (Fig. 8.2).

The third (SDG3), to "ensure healthy lives and promote well-being for all at all age" is specific to health (Fig. 8.3).

SDG3 includes the provision of universal health coverage (UHC; SDG3.8), which aims to provide access to good-quality health services for all, without financial hardship. SDG3.3 is targeted at ending the prevalence of neglected tropical diseases. But, multisectoral action to address poverty, control disease vectors and the environment, and improve access to clean water and sanitation are key components of neglected tropical disease programs.

FIG. 8.2

Sustainable Development Goals (17 goals) to be completed by the end of 2030.

FIG. 8.3

SDG3 showing healthy lives and promote well-being for all ages.

SDG 3 consists of 13 targets and 28 indicators to have accountability of progress. The first nine targets are known as "outcome targets" which include reduction in maternal mortality; completely stop of maternal death under 5 years of age; prevention of communicable diseases; ensure reduction of mortality from noncommunicable diseases and promote mental health; prevent and treat substance abuse; reduce road injuries and deaths; sexual and reproductive care, family planning and education; achieve universal health coverage; and reduce illness and deaths from hazardous chemicals and pollution.

The four "means to achieve" SDG 3 targets include implement the WHO Framework Convention on Tobacco Control; support research and development of affordable vaccines and medicines; support financially for the health workforce in developing countries; and improve early warning systems for global health risk.

Additionally, SDG 3 targets to successfully implement and achieve universal health coverage, which is means for equitable access of healthcare services to all people so that it could be helpful to end the preventable death of newborns, infants, and children under the age group of 5 years.

Other SDGs (e.g., on hunger, gender equality, clean water, and sanitation, affordable and clean energy; sustainable cities and communities; climatic action; and peace, justice, and strong institutions) of health also indirectly support PHC.

All people, irrespective of caste, race, nation, everywhere, deserve the right care, right in their community, which is basic to the premise of primary health care. Primary health care (PHC) is essential to a person for leading sound health throughout the life. This includes physical, mental, and social well-being of all people at all times. It should cover health promotion, disease prevention, treatment, rehabilitation, and palliative care. The best way to approach the people for PHC is to meet their health needs throughout their lives. In order to bring awareness on broader determinants of health through the multispectral policy, it is necessary to convince the population at individual, family, and community levels to take responsibility for their own.

In 1978, International Conference on Primary Health Care was held in Alma Ata, Kazakhstan and became a core concept of the World Health Organization's goal of Health for all. The Alma-Ata conference mobilized a "Primary Health Care Movement" of professionals and institutions, governments and civil society organizations, researchers, and grassroot organizations that undertook to tackle the "policy, socially and economically unacceptable" health inequalities in all countries.

After a gap of four decades, global leaders could realize the importance of the primary health-care system and brought few amendments in the Declaration of Astana summit in order to make it globally acceptable.

In 2005, the United Progressive Alliance Government launched the National Rural Health the Mission (NRHM) to improve access to quality health care, especially for poor rural women and children, to upgrade primary health-care institutions, increase equity, and the decentralization of services, and encourage states to generate alternate sources of financing.

8.3 Alma-Ata summit

In September 1978, Soviet Socialist Republic held an international conference on primary health care at Almaty (formerly Alma-Aata), with the need for urgent action by all government workers, and the world community to protect and promote the health of all people. This was the first international underlining on the social significance of primary health care. The other primary members of the World Health Organization (WHO) also accepted the norms of Alma. So, from the global point of view, the main motto of the conference is to collaborate in introducing, developing, and maintaining primary health care for all the people. The declarations have 10 points and are nonbinding on member states:

The first section is about Alma-Ata acceptance of definition on primary health care as defined by the WHO: "a state of complete physical, mental, and social well-being and not merely the absence of disease or infirmity," The definition includes social and economic sectors within the scope of attaining health and reaffirms health as a human right.

The second section explains that inequality of health status between the developed and developing countries is absurd and not an acceptable term, politically, socially, and economically.

The third section reveals how primary health care and economic and social development are complement to each other. Concomitantly, it is responsible for bringing world peace through the promotion and protection of the health of the people. Participation of people as a group or individual in planning and implementing their health care was declared as a human right and duty.

The fourth section emphasized the responsibility and commitment of the member states in providing adequate health and social measures. As declared by WHO "Health for All," all the member states are supposed to campaign for universal health coverage. The declaration also urged governments, international organizations, and the global community to take this challenge as a social target in the spirit of social justice.

The fifth section explains the responsibility of the Sovereign state in providing sufficient health and social measures for bringing awareness on primary health care and support WHO's call for "Health for All."

The sixth section is all about the benefit of the primary health-care system as Margret Chan the Director-General of the WHO has reaffirmed that the primary health-care approach is the most efficient and cost-effective way to organize the health system. Margret Chan is a Chinese Canadian physician, who served as the Director-General of the WHO delegating the People's Republic of China during the period 2000–2017. She also highlighted those results on primary health care at the global level, which were highly encouraging due to the low cost with higher satisfaction.

The seventh section is about the components of primary health care. The subsequent two sections are for the governments to implement a primary health-care approach in their health systems, which urged international cooperation for the better use of the world's resources.

After about 5 years all other countries have accepted the Alma-Alta Declaration which has emerged as a major milestone of the 20th century for primary health care. Meanwhile, Communist China has started paying keen attention to PHC to serve the rural population at the community level massively. The "barefoot doctors" played a critical role to support the primary health-care campaign in the rural part of China.

Barefoot doctors are health-care providers who undergo basic medical training and worked in rural villages in China. They included farmers, folk healers, rural health-care providers, and recent middle or secondary school graduates who received minimal basic medical and paramedical education. Their main reason for bringing health care to rural areas is the urban-trained doctors would not prefer to settle in rural areas.

Subsequent to the Alma-Ata declaration, WHO, UNICEF, and other international organizations, as well as multilateral and bilateral agencies, nongovernmental organizations, funding agencies, all health workers, and world comity started showing interest for global approach in implementing primary health care and to channel increaser financial and technical, particularly in developing countries.

8.4 Criticism on Alma-Ata

The Alta-Ata Declaration is under a wide range of disputes due to its impracticable work plan without any time target. The target for "health for all by the years 2000" was planned without any guarantee. In this connection, the Rockefeller Foundation in 1979 held a conference in order to develop a most cost-effective blueprint to understand the status of interrelation between health and population programs.

In response to the vagueness of primary health care and the declaration of Alma Ata, an alternative movement began to gain momentum. After a year of Alma-Ata declaration, Julia Walsh and Kenneth Warren suggested bringing certain amendments as "Special Primary Health care (SPHC) [9]. The main base for such suggestions is to fight disease based on cost-effective medical intervention. Although they acknowledge that the goal set at Alma Ata was admirable.

The main significance of SPHC is to focus on the most severe public health problem. By narrowing the target, the quality of health care can be successfully managed within the target time. For example, SPHC identified four factors to guide the selection of target diseases for prevention and treatment: prevalence, morbidity, mortality, and feasibility of control. In this connection, SPHC highlighted four vertical programs: growth monitoring, oral rehydration therapy, breastfeeding, and immunization (GOBI) [10]. At a later stage, family planning, female education, and food supplementation (FFF) were added. The advantage of such additional programs is mainly due to their easy measurable fact [11]. This would be helpful for easy funding and developing successful and authentic reports.

8.5 Medicine vs public health

There have been recurring issues between medicine (treatment) and public health (prevention). This is mainly due to heavy competition on the development of value-added medicines by most modernized pharmaceutical companies. Most of the medicines have been manufactured under the strict guidelines of international monitoring agencies, in order to keep safety and security.

The continuous therapeutic revolution in drug design and development is mainly due to the maintenance of biological superiority of drugs from the quality point of view. For example, Penicillin is a miracle drug that not only cured illnesses but also saved entire societies from political and economic collapse [12].

Medicines are for maintenance of health care at different stages of the life cycle. Irrespective of cost, medicines are used at primary, secondary, and territory levels of health treatment. Contemporary medicine applies biomedical sciences, biomedical research to diagnose, treat, and prevent injury and disease. But, due to the cost and market availability, it is not possible to use many lifesaving biologics.

Availability of medicines and clinical practice varies across different geographic locations of the world, and the availability and cost of medicines vary accordingly. Biologic drugs are highly developed in the Western world. While in developing countries such as Asia-pacific and Africa, the population relies on traditional pharmaceutical medicines due to their low cost.

The provision of medical care for public health is classified into primary, secondary, and tertiary care categories.

Primary health-care medical services are extended by physicians, physician assistants, nurses, or other health professionals who have the first contact with a patient seeking medical treatment or care. About 90% of medical visits can be treated by the primary care provider. These include treatment of acute and chronic illnesses, preventive care, and health education for all ages and both sexes.

Secondary care medical services are provided by medical specialists in their offices or clinics or local community hospital for a patient referred by a primary care provider who first diagnosed or treated the patient. Some primary care providers may also take care of hospitalized patients and deliver babies in a secondary care setting.

Tertiary care medical services are provided by specialized hospitals or regional health center well equipped with diagnostic and treatment facilities. These include trauma center, burn treatment center, advanced neonatology unit services, organ transplants, high-risk pregnancy, and radiation oncology, etc.

Modern medical care is also well provided by information technology devices for recording and communicating information in the shortest possible time.

In low-income countries, modern health care is often too expensive for the average person. International health-care policy researchers have advocated that "user fees" be removed in these areas to ensure access, although even after removal, significant costs and barriers remain.

8.6 Components

There are eight essential components of PHC [13], including [1] health education, on prevailing health problems and the methods of preventing and controlling them, [2] nutritional promotion including food supply, [3] supply of adequate safe water and sanitization, [4] maternal and child health care, [5] immunization against major infectious diseases, [6] prevention and control of locally endemic diseases, [7] appropriate treatment of common diseases and injuries and [8] provisions for essential drugs, all these basic requirements are incorporated in the SDGs for 2030 from goal 2 to goal 4 [14].

Since the last four decades, PHC Provision has moved from simple planning to action. Alma-Ata Declaration is still crucial in current global health like COVID-19 pandemic, especially for developing countries.

To achieve this, initially, the United Nations (UN) announced eight Millennium Development Goals (MDGs) by 2015. But looking at the progress, the UN announced 17 Sustainable Development Goals (SDGs), with the strong hope to achieve the same by the end of 2030.

The third (SDG3) to "ensure healthy lives and promote well-being for all at all age" is specific to health. SDG3 includes the provision of universal health coverage (UHC; SDG 3.8), which aims to provide access to good-quality health services for all, without financial hardship. Other SDGs (e.g., on hunger, gender equality, clean water and sanitation, affordable and clean energy; sustainable cities and communities; climatic action; and peace, justice, and strong institutions) of health also indirectly support PHC. Despite developing innovative sustainable models on PHC, still, it has been tough to maintain quality PHC services in isolated communities having a small and randomly distributed population [15, 16].

8.7 Pillars of primary health care

The primary health-care system is the key factor for community improvement and balance socioeconomic conditions. It reduces the inequalities between different groups of a community. The primary health care is basic health care with applied,

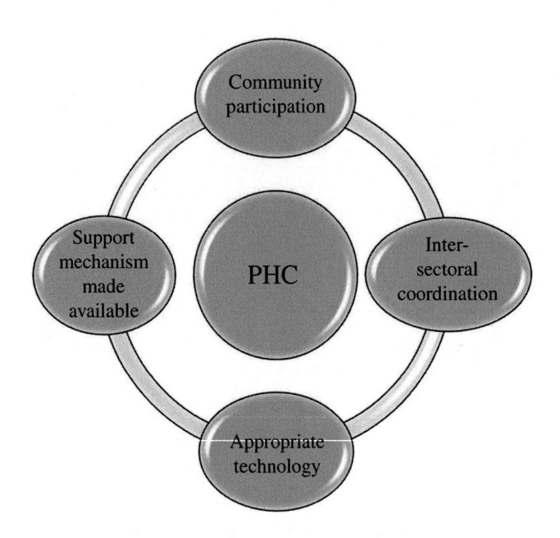

FIG. 8.4

The four major pillars of the primary health-care system.

scientifically sound, and universally acceptable methods and technology, which should be available and accessible to all individuals and families in a community, on a priority basis.

Successful primary health care can only be possible through an integrated system being coordinated by different workforces who are well-trained in health-care management. The primary health-care outline is built on four key pillars (Fig. 8.4).

The four major pillars of primary health care that should be made available are as follows: community participation, intersectorial coordination, appropriate technologies, and support mechanisms.

8.7.1 Community participation

- Community participation is a process in which community people voluntarily serve their community health care.
- It is absolutely a social approach.
- The participants should identify the health needs of the community, planning, organizing, decision-making, and implementation of health programs, sponsored by the government or NGO.
- It also ensures effective and strategic planning and evaluation of health-care services.

8.7.2 Intersectoral coordination

- Intersectoral coordination plays a significant role in performing different aspects of health-care services.
- The involvement of specialized NGOs, private sectors, and public sectors is important for the successful operation of the health-care system.

- It also refers to delivering health-care services in an integrated way.
- It is necessary that departments such as agriculture, animal husbandry, food industry, education, housing, public works, communication, and other sectors need to be involved in achieving health for all.

8.7.3 Appropriate technologies

- The technologies to be implemented for health-care system development should be available and accessible for health-care services.
- The technologies should be scientifically sound, adaptable to local needs, and acceptable to the doctor and health-care workforces.

8.7.4 Support mechanisms made available

- Support mechanisms are vital to health and quality of life.
- Support mechanism includes that the people are getting personal, physical, mental, spiritual, and instrumental support to meet goals of primary health care.
- Primary health care depends on an adequate number and distribution of trained physicians, nurses, community health workers, and others working as a health team and supported at the local and referral levels.

8.8 Salient features of primary health care

- Primary health care is mainly based on quality and cost effectiveness.
- Primary health care highlights "Health for All".
- Primary health-care system integrates preventive, promotive, curative, rehabilitative, and palliative health-care services.
- Primary health care promotes social inclusion: It includes services that are readily accessible and available to the community.
- Primary health care can be easily available, even at the time of emergency caused due to natural calamity and biological disaster.
- Primary health-care promotes equity and equality.
- Primary health-care improves safety, performance, and accountability.
- Primary health-care pleads on health promotion and focuses on prevention, screening, and early intervention of health disparities.
- Primary health care also promotes socioeconomic condition of a community.

8.9 Challenges for implementation of PHC

In order to meet the challenges of primary health care, it is necessary to share ideas and experiences, test new approaches, and share the results to enable successful innovations to spread more rapidly between organizations and across geographies.

The following challenges can establish and scale primary care innovations and the solutions to overcome them:

- shortage of well-qualified physicians and health-care workforces to serve the patients in emergency,
- inadequate technology and equipment,
- poor condition of infrastructure, especially in rural areas,
- concentrated focus on curative health services rather than preventive and promotive health-care services,
- challenging geographic distribution,
- poor quality health-care services,
- inadequate financial support in health-care programs,
- lack of community participation,
- poor allotment of health-care workforce to serve in the rural locality, and
- lack of intersectoral collaboration.

8.10 Role of primary care in the COVID-19 response

Massive health disruption has been occurring since the outbreak of COVID-19 from epidemic to pandemic. Cases of casualty due to COVID-19 all over the world are alarming [17].

There are huge differences between countries, the burden of COVID-19 on societies and economies, and the measures being implemented globally. The majority of the people infected with COVID-19 have a self-limiting infection and recovery. A minority of the population with 10% of cases require intensive care unit admission. Unfortunately, some patients pass away (Table 8.2).

Although all age groups are at risk of contracting COVID-19, older people face a significant risk of developing severe illness if they contract the disease due to physiological changes that come with aging and potential underlying health conditions. Presently, the severity of COVID-19 infection is noticed more in older adults. For example, in the European region, the top 30 countries are with the largest percentage of older people suffering from COVID-19.

Presently, people from all age groups are susceptible to COVID-19, but elderly people, above the age of 60 are more prone to COVID-19 due to physical changes that come with aging and potential underlying health condition. The pandemic nature of COVID-19 has completely changed the regular lifestyle of older people, and also the support they used to receive from the community. Prolonged stay of elderly people at home challenges their physical ability and mental condition. So, it is high time that we create opportunities to foster healthy aging during the pandemic. In this connection, WHO, in collaboration with other partners, is providing guidance and advice during the COVID-19 pandemic for older people and their householders, health and social care workers and local authorities, and community.

Table 8.2 COVID-19 situation update for the EU/EEA, as of June 2, 2021.

Country	Cases	Deaths	Date of data collection
Austria	641,044	10,355	02/06/2021
Belgium	1,063,405	24,968	01/06/2021
Bulgaria	418,813	17,726	02/06/2021
Croatia	356,397	8034	02/06/2021
Cyprus	72,515	360	02/06/2021
Czechia	1,662,256	30,126	02/06/2021
Denmark	282,135	2516	02/06/2021
Estonia	129,804	1259	02/06/2021
Finland	92,642	956	02/06/2021
France	5,677,172	109,691	02/06/2021
Germany	3,687,828	88,774	02/06/2021
Greece	404,163	12,122	02/06/2021
Hungary	804,987	29,774	02/06/2021
Iceland	6590	30	02/06/2021
Ireland	261,517	4941	02/06/2021
Italy	4,220,304	126,221	02/06/2021
Latvia	133,518	2379	02/06/2021
Liechtenstein	3016	58	02/06/2021
Lithuania	275,198	4283	02/06/2021
Luxembourg	69,983	817	02/06/2021
Malta	30,543	419	02/06/2021
Netherlands	1,649,646	17,610	02/06/2021
Norway	125,071	783	02/06/2021
Poland	2,872,868	73,856	02/06/2021
Portugal	849,538	17,025	02/06/2021
Romania	1,077,978	30,353	02/06/2021
Slovakia	774,919	12,353	02/06/2021
Slovenia	254,045	4694	02/06/2021
Spain	3,682,778	79,983	01/06/2021
Sweden	1,068,473	14,451	26/05/2021

Source: European Centre for Disease Prevention and Control, an agency of the European Union.

The United Nations General Assembly declared 2021–2030 the Decade of Healthy Aging. The Decade of Healthy Aging 2020–2030 is a global collaboration, aligned with the last 10 years of the Sustainable Development Goals (SDGs) that brings together government, civil society, nongovernmental Organizations (NGOs), international agencies, professionals, academia, the media, and the private sector to improve the lives of older people, their families, and the communities in which they live.

The process of early aging is at a faster rate due to environmental changes and climatic changes at a faster pace than in the past. The demographic transition catalyzes the country's economic development. The demographic transition refers to population trends of two demographic characteristics: birth rate and death rate to suggest that a country's total population growth rate cycles through stages as that country develop economically. So, it is obvious that world population is aging at a faster pace than in the past. The demographic transition will have an impact on almost all aspects of society.

Although all age groups are at risk of contracting COVID-19, older people face a significant risk of developing severe illness if they contract the disease due to physiological changes that come with aging and potential underlying health conditions.

At present more than 1 billion people in their 60s are from low- and middle-income countries. Many of them have inadequate access to even the basic resources necessary for a sustainable healthy life. The alarming Covid-19 has challenged the gaps in policies, systems, and services. Under the pandemic condition of COVID-19, it is necessary to plan on how to save the older people and their families and communities from the grip of COVID-19 and develop an ideal strategy for the continuity of healthy life, till the pandemic is completely eradicated. It is high time to mitigate the massive damages caused by the COVID-19 pandemic, the need to maintain and improve essential primary health-care services on a priority basis. This situation underscores the importance of the primary health-care revitalization agenda articulated in the 2018 Astana Declaration.

Regulation and control of health are not only by genetic inheritance but also closely related to the surrounding social- and natural environment where we live in. Our physical, mental, and body function throughout the life cycle, and especially in old age has to adjust with the environment, despite the loss of body structure and function with the passing of time. In this connection, the long-term effect of COVID-19 is still subject to much research. But, it has been in the process of understanding that the severity of COVID-19 may cause harm to the immune system.

Environments play an important role in determining our physical and mental capacity across a person's life course and into older age and also how well we adjust to the loss of function and other forms of adversity that we may experience at different stages of life, and in particular, in later years. Both older people and the environments in which they live are diverse, dynamic, and changing. In interaction with each other, they hold incredible potential for enabling or constraining *Healthy Aging*.

It reemphasizes the importance of primary health care to address current health challenges, renewing political commitment to primary health care, and achieving universal health coverage.

8.11 Astana declaration on primary health care

In 2018, after a gap of four decades of Alma-Ata declaration on primary health care, world leaders, government ministers, development partners, civil society, and young people organized a conference on primary health care. Astana conference was jointly hosted by the Government of Kazakhstan, UNICEF, and WHO.

The Astana declaration is based on "Commitment to the fundamental right of every human being to the enjoyment of the highest attainable standard of health without distinction of any kind." The Astana declaration reemphasizes the importance of primary health care to address current health care challenges, renewing political commitment to primary health care, and achieving universal health coverage with a special reference to the rural community.

World Health Organization also endorsed the Astana decoration for a newly amended frame for primary health care with special emphasis on the rural community. The four key areas are as follows: (i) make bold political choices for health across all sectors; (ii) build sustainable primary health care; (iii) empower individuals and communities; and (iv) align stakeholder support to national policies, strategies, and plans.

(i) *First Astana declaration*

The first declaration refers to "make bold political choices for health across all sectors." The basic theme of the declaration is to promote multisectoral action through the health care system, as per the WHO protocol [1].

The main target is to convince the stakeholders about the current challenges in primary health care for the complete and successful implementation of health-care services, as per schedule. At the international forum, there are policies and toolkits to handle conflicts of interest; for example, the National Health System (NHS) in England [18].

The governments at all levels should protect the right of the each and every person to get the maximum privilege of health services. In addition, emphasis is also given to promote multisectoral action and UHC, involving relevant stakeholders and empowering local communities to strengthen PHC. The Astana declaration also mentioned that the government policy decision should consider economic, social, and environmental determinants of health at the rural community level.

(ii) *Build sustainable primary health care*

The second Astana declaration is to "build sustainable primary health care." Primary health-care service should not be targeted to a simple implementation of policy but the emphasis is to be given on how efficiently it is to be worked out to bring sustainability in health service. For example, the Australian health-care reform agenda is based on high priority to integrated, compressive PHC services that are sustainable and responsive to community needs [19].

(iii) *Empower individuals and communities*

The third commitment of the new declaration is to "empower individuals and communities." The basic intentions of the third commandment are community involvement, public participation, and empowerment and health literacy. In addition, Astana post declaration includes increasing people's knowledge on health maintenance through affective camping on health management awareness.

The European Patients' Academy on Therapeutic Innovation is an example of an institution established to bring awareness among patients through education.

(iv)*Align stakeholder support to national policies, strategies, and plans*

Align stakeholders to support national policies, strategies, and plans related to health services, especially in rural communities. The stakeholders include health professionals, academia, patients, civil society, local and international partners, agencies and funds, the private sector, and NGOs. Stakeholder support can be helpful for developing sufficient well-trained health professions, health care-related technologies easily acceptable to physicians and other health-care workforce, and financial and information resources to PHC.

8.12 Overall target to upgrade PHC for COVID-19 control

The overall target of the Astana declaration is to revitalize the entire decision taken on primary health care in past and strength PHC by prioritizing disease prevention and health promotion and aims to meet all people's health needs across the life course through compressive preventive, promotive, curative, rehabilitation services, and palliative care. So, it is high time to find out comprehensive data survey report on COVID-19 and bring stagnation in the further progress of pandemic situation with the guidelines of 2020 World Health Organization Operational "framework for primary health care."

In addition, to strengthen the present scenario of PHC, global level initiative requires to have control over COVD-19. This needs political commitment and leadership, governance and policy, funding and allocation of resources, and engagement of communities and other stakeholders. For example, India, the world's second-highest populated country, with the guideline of WHO could launch the largest COVID-19 vaccination drives in the world. The WHO Country Office for India (WCO India) has been closely associated with the COVID-19 eradication drive linked to surveillance and contact tracing, clinical diagnosis, emergency health workforce management, infection prevention and control, timely vaccination, etc. at national, state, and district levels. At the national level, WCO India is actively involved with the Ministry of Health and Family Welfare (MoHFW) in collaboration with the Joint Monitoring Group (JMG) and National Center for Disease Control (NCDC), Indian Council of Medical Research (ICNR), National Disaster Management Authority, and NITY Aayog. Additionally, WHO teams are also closely associated with the National and State Governments in primary health services like immunization, reproductive, maternal, newborn child, and adolescence health (RMNCAH), and noncommunicable diseases prevention and control. So, the COVID-19 controlling system in India can be a good model to control the pandemic, under various adverse conditions.

8.12.1 How to bring sustainability

Challenging guarantee healthy and progressive well-being at all ages is indispensable to sustainable development. At present, the world is in the grip of the pandemic COVID-19 and bringing disaster in public life with complete disruption of the global economy, and dismissing the life of billions of people all around the world. The PHC service is the initial phase to control contagious communicable disease. Any lacunas leftover in PHC may lead to an epidemic followed by a pandemic. So, it is necessary to strengthen PHC in order to bring sustainability to the health system, especially in rural and remote areas. It is obvious that changes in nature are supposed to occur (biological or natural disasters) during a pandemic situation. But we have to find out possible controlling strategies which include workforce supply, to find a way-out to send health workforce professionals to a rural and remote community, integrate the stakeholders for timely cooperation in PHC operation, take care of noncommunicable diseases, especially of elderly people (because rising chronic disease burden may increase the number of causality in pandemic condition), temporary infrastructure demand, and leadership and management accountability.

Sustainability in PHC is the primary need of the day for each and every body, irrespective of nationality. There are a lot of differences that exist between countries in the organization of primary health care and availability of human resources. In SDG3, challenges to overcome such problems (related to reproductivity and child health, communicable diseases, chronic illness, addiction, and other mental health problems) are explained clearly how effectively one can go ahead with the population-based approach to primary health care [19–23]. In addition, it is also explained that the strategies for delivering vaccines and drugs need a functioning primary care system. So, during pandemic situations well-integrated and prepared primary health care has a key role in health emergency responsiveness, and it is essential for the achievement of equitable and cost-effective UHC [24–26].

Basically, sustainability means the ability of a health service to provide ongoing access to appropriate quality care in a cost-efficient and health-efficient manner. In order to bring sustainability in health care, 17 sustainable development goals (SDGs) were adopted at the 70th session of the United Nations (UN) General Assembly in 2015. Detail, in this connection, has already been described in the earlier section of this chapter (Section 8.1). It is also necessary to implement other health-related SDGs like SDH-2 on the basis of a multisectoral approach. For this purpose, we need strong governance. For example, to end hunger (SDG2), primary health care can support community-based therapeutic care using ready-to-use therapeutic foods for severe acute malnutrition.

For a successful implementation of SDH 11 and development of sustainable cities and communities, primary health care can support the monitoring of air pollution, issue health warnings when the particulate matter of diameter less than 2.5 μm in the air exceeds limits, and advocate for the reduction of indoor pollution through the use of clean energy. All this could possible through policies that address industrial, residential, and vehicular sources of pollution. But, still, the rural areas and remote areas of developing and underdeveloped countries are lacking basic PHC services mainly due to extremely poor transportation and other communication facilities; lack of economies of scale; difficulty in maintaining an adequate workforce, poor management structures, and geographic isolation. So, people suffer from acute and chronic diseases. So, it is necessary for a government to increase its financial commitment to the health sector. It is true that a sound health system is a complement to sustainable development. In this connection, we should collaborate and focus on broader economic and social inequalities, urbanization, climate crisis, continuing burden of HIV and other infectious diseases, and non- communicable diseases, as prescribed in SDGs.

References

[1] Bhore Committee, https://pgblazer.com/bhore-committee-1946/?_e_pi_=7%2CPAGE_ID10%2C4365817816; 1946.
[2] Newell KW. Health by the people. Geneva: World Health Organization; 1975. p. xi.
[3] Canadian Department of National Health and Welfare. A new perspective on the health of Canadians. Ottawa; 1974.
[4] Tangcharoensathien V, Mills A, Palu T. Accelerating health equity: the key role of universal health coverage in the sustainable development goals. BMC Med 2015;13(1):101. https://doi.org/10.1186/s12916-015-0342-3.
[5] International Conference on Primary Health Care. Declaration of Alma-Ata. WHO Chron 1978;32(11):428–30.
[6] A vision for primary health care in the 21st century: towards universal health coverage and the Sustainable Development Goals. Geneva: World Health Organization, United Nations Children's Fund; 2018. Available from: https://apps.who.int/iris/handle/10665/328065.
[7] Hone T, Macinko J, Millett C. Revisiting Alma-Ata: what is the role of primary health care in achieving the sustainable development goals? Lancet 2018;392(10156):1461–72. https://doi.org/10.1016/S0140-6736(18)31829-4.
[8] Starfield B. Politics, primary healthcare and health: was Virchow right? J Epidemiol Community Health 2011;65(8):653–5. https://doi.org/10.1136/jech.2009.102780.
[9] Walsh J, Warren K. Selective primary health care: an interim strategy for disease control in developing countries. N Engl J Med 1979;301(18):967–74.
[10] Ruxin J. Magical bullet: the history of oral rehydration therapy. Med Hist 1991;38:363–97.
[11] Cueto M. The origins of primary health care and selective primary health care. Am J Public Health 2004;94:1864–74.
[12] Bud R. Penicillin, triumph and tragedy. New York: Oxford University Press; 2009. p. 75–112.
[13] World Health Organization. Primary health care: Report of the international conference on primary health care, Alma-Ata, USSR. Geneva: World Health Organization; 1978.
[14] United Nations. Transforming our world: the 2030 agenda for sustainable development, https://sustainabledevelopment.un.org/post2015/transformingourworld/publication.
[15] Battye KM, McTaggert K. Development of a model for sustainable delivery of outreach allied health services to remote north-west Queensland, Australia. Rural Remote Health 2003;3(3):194.
[16] Hoodless M, Evans F. The multipurpose service program: the best option for rural Australia. Aust J Prim Health 2001;7(1):90–6. https://doi.org/10.1071/PY01015.
[17] World Health Organisation. WHO Coronavirus (COVID-19) Dashboard, http://Covid19.who.int; 2021.
[18] Mahler H. Primary health care comes full circle. Bull World Health Organ 2008;86:747–8.

[19] Bhutta ZA, Ali S, Cousens S, et al. Interventions to address maternal, newborn, and child survival: what difference can integrated primary health care strategies make? Lancet 2008;372:972–89.
[20] Beaglehole R, Epping-Jordan J, Patel V, et al. Improving the prevention and management of chronic disease in low-income and middle-income countries: a priority for primary health care. Lancet 2008;372:940–9.
[21] Kaner E, Beyer F, Dickinson H, et al. Effectiveness of brief alcohol interventions in primary care populations. Cochrane Database Syst Rev 2007;2, CD004148.
[22] Kruk ME, Porignon D, Rockers PC, Van Lerberghe W. The contribution of primary care to health and health systems in low- and middle-income countries: a critical review of major primary care initiatives. Soc Sci Med 2010;70:904–11.
[23] Gibbons DC, Bindman AB, Soljak MA, Millett C, Majeed A. Defining primary care sensitive conditions: a necessity for effective primary care delivery? J R Soc Med 2012;105:422–8.
[24] Lewin S, Lavis JN, Oxman AD, et al. Supporting the delivery of cost-effective interventions in primary health-care systems in low-income and middle-income countries: an overview of systematic reviews. Lancet 2008;372:928–39.
[25] Starfield B, Shi L, Macinko J. Contribution of primary care to health systems and health. Milbank Q 2005;83:457–502.
[26] Rao M, Pilot E. The missing link- the role of primary care in global health. Glob Health Action 2014;7:23693.

Index

Note: Page numbers followed by *f* indicate figures and *t* indicate tables.

Printed in the United States
by Baker & Taylor Publisher Services